AF800626

The Future of EU Agricultural Markets
by AGMEMOD

Frédéric Chantreuil • Kevin F. Hanrahan
Myrna van Leeuwen
Editors

The Future of EU Agricultural Markets by AGMEMOD

Springer

Agri-food projections
AGMEMOD
for EU member states

Editors
Dr. Frédéric Chantreuil
Université de Caen Basse Normandie
UFR Sciences Economiques et Gestion
17 rue Claude Bloch, BP 5186,
F-14032 Caen Cedex
France
Frederic.Chantreuil@unicaen.fr

Myrna van Leeuwen
Research Area International Policy
LEI Wageningen UR
Alexanderveld 5
2585 DB The Hague
The Netherlands
Myrna.vanLeeuwen@wur.nl

Dr. Kevin F. Hanrahan
Teagasc
Rural Economy Development Programme
Agricultural Economics and Farm Survey
Department
Athenry
Galway
Ireland
Kevin.Hanrahan@teagasc.ie

Files, programs and tools of the AGMEMOD 4.0 version are available for download at Springer's Extra Materials website: http://extras.springer.com/

ISBN 978-94-007-2290-3 e-ISBN 978-94-007-2291-0
DOI 10.1007/978-94-007-2291-0
Springer Dordrecht Heidelberg London New York

Library of Congress Control Number: 2011940680

© Springer Science+Business Media B.V. 2012
No part of this work may be reproduced, stored in a retrieval system, or transmitted in any form or by any means, electronic, mechanical, photocopying, microfilming, recording or otherwise, without written permission from the Publisher, with the exception of any material supplied specifically for the purpose of being entered and executed on a computer system, for exclusive use by the purchaser of the work.

Printed on acid-free paper

Springer is part of Springer Science+Business Media (www.springer.com)

Acknowledgements

We would like to acknowledge the members of the AGMEMOD Advisory Group – Loek Boonekamp (OECD), Lauri Kettunen (MTT), Hans-Jörg Lutzeyer (European Commission-DG RTD), Wolfgang Münch (European Commission-DG AGRI), Emilio Rossi (Global Insight), Patrick Westhoff (FAPRI at the University of Missouri) – for the fruitful comments, suggestions and advice they have provided to the Partnership since the foundation of the AGMEMOD project.

The authors would like to thank the Institute for Prospective Technological Studies (IPTS) and DG RTD for their support and assistance in jointly organising validation workshops hosted at IPTS Seville and at DG AGRI Brussels. We would like to thank representatives of DG AGRI, DG JRC, DG RTD, FAPRI, OECD, FAO, University of Bonn, University of Gottingen and University of Aberdeen for the helpful criticisms, remarks and suggestions they made concerning the AGMEMOD results that were presented during these workshops.

Any opinions and findings, conclusions, suggestions or recommendations expressed in this book are those of the authors and do not necessary reflect the view of European Commission Services or Research Institutes and Universities who are supporting the AGMEMOD Partnership.

F. Chantreuil
K.F. Hanrahan
M. van Leeuwen

Contents

AGMEMOD Partnership

Universität für Bodenkultur Wien	Martin Kniepert, Markus Hofreither
Université Catholique de Louvain	Bruno Henry de Frahan, Henrich Brunke
Institute of Agriculture Economics	Nedka Ivanova, Plamen Mishev, Mariya Peneva
Institute of Agricultural Economics and Information	Ivan Foltyn, Jan Kubat, Ida Zednickova
Food and Resource Economic Institute	Jørgen D. Jensen, Lars-Bo Jacobsen
Institute of Economics and Social Sciences Estonian Agricultural University	Mati Sepp, Tönu Akkel
MTT Agrifood Research Finland	Jyrki Niemi, Heikki Lehtonen, Lauri Kettunen
Institut National de la Recherche Agronomique	Frédéric Chantreuil, Marie-Dominique Le Barbenchon, Fabrice Levert
Johann Heinrich von Thünen-Institut, Federal Research Institute for Rural Areas, Forestry and Fisheries	Petra Salamon, Oliver von Ledebur
Department of Economics, University of Athens	George Mergos, Anastasios Portokalakis, Elias Mantzouneas
Corvinus University of Budapest	Tibor Ferenczi, Tibor Varga

Rural Economy Research Centre, Teagasc	Trevor Donnellan, Kevin F. Hanrahan, Peter Howley
Università Politecnica delle Marche	Roberto Esposti, Andrea Bonfiglio, Antonello Lobianco, Giulia Listorti, Beatrice Camaioni
Latvian State Institute of Agrarian Economics	Guna Salputra, Andris Miglavs, Andrejs Krotovs
Lithuanian Institute of Agrarian Economics	Irena Kriščiukaitienė, William H. Meyers, Salomeja Andrikiene, Aistė Galnaitytė, Andrej Jedik
LEI Wageningen UR	Myrna van Leeuwen, Andrzej Tabeau, Wietse Dol, Foppe Bouma
Warsaw Agricultural University	Edward Majewski, Stanisław Stańko, Mariusz Hamulczuk, Katarzyna Kowalska, Sylwia Krawczyńska
Universidade Nova de Lisboa	Fernando Brito Soares, Tiago Silva Vieira
Institute of Agricultural Economics	Dinu Gavrilescu, Camelia Gavrilescu, Cristian Kevorchian
Slovak Agricultural University	Jan Pokrivcak, Dusan Drabik, Veronika Nahacka, Pavel Ciaian
University of Ljubljana, Biotechnical Faculty	Emil Erjavec, Stane Kavčič, Darja Regoršek, Marjeta Pintar, Barbara Zagorc
Unidad de Economia Agraria, Zaragoza	Azucena Gracia, Tiziana de Magistris
Queen's University, Belfast	Ziping Wu, Jos Ijpelaar, Lei Qu
DG JRC-IPTS-EC, Seville	Robert M'barek

List of Boxes

List of Figures

List of Tables

Part I

Introduction

The purpose of this chapter is to give a brief outline of respectively the AGMEMOD model, the purpose of this book, the structure of the book and its usefulness to the reader.

What is AGMEMOD?

AGMEMOD stands for **Ag**ricultural **Me**mber State **Mod**elling and was established in 2001. Originally coordinated by Brendan Riordan of Teagasc Ireland, the AGMEMOD Partnership comprised universities, research institutes and government agencies from EU15 Member States. In 2002 the Partnership was extended to include partners from the countries that acceded to the European Union (EU) in May 2004 and in January 2007. Since 2007, partners from EU Candidate Countries (Macedonia, Croatia and Turkey) and other European countries (Russia, Ukraine, and Kazakhstan) have joined the AGMEMOD Partnership. All groups that become members of the AGMEMOD Partnership sign an agreement concerning the common ownership of the models developed and the analytic results.

AGMEMOD was funded under the European Commission 5th and 6th Framework Programmes (respectively QLRT-2001-02853 and SSPE-CT-2005-021543) and by contributions from the partners' institutes throughout the EU. The development of the AGMEMOD model's analytic capacity was also supported by projects funded by the Institute for Prospective Technological Studies (IPTS), part of the European Commission's Joint Research Centre (JRC). While the coordination for the FP6 project was being provided by INRA, France, the task of managing short to medium-term projects within the AGMEMOD partnership became the responsibility of LEI, the Netherlands.

AGMEMOD is an econometric, dynamic, multi-product partial equilibrium model wherein a bottom-up approach is used. Based on a set of commodity specific model templates, country specific models were developed to reflect the detail of agriculture at Member State level and at the same time to allow for their combination in an EU model. This approach allows the inherent heterogeneity of the agricultural systems existing across the EU to be captured within the model's parameterisation, while the analytical consistency across the country models is ensured through the adherence to agreed commodity model templates. The maintenance of analytical consistency across the country models is essential for the successful aggregation of country models to the EU level. It also facilitates the meaningful comparison of the impact of a policy change across different Member States.

Why AGMEMOD?

The primary objective of the AMEMOD Partnership is to develop and maintain a partial equilibrium modelling system with the capacity to undertake model-based economic analysis of the impact of policy or other changes on the agri-food sector of each EU Member State and the EU as a whole.

The development, ongoing maintenance and improvement of the AGMEMOD model mark an advance in agricultural sector model building research as up until now the building and use of multi-country models for Europe's agri-food sector has been done in one institution rather than in each of the modelled countries as in the AGMEMOD project. The AGMEMOD Partnership's approach, wherein a bottom-up approach is used, is based on the development of country level models to a common country model template and their subsequent combination in a composite EU model. This approach seeks to better capture the inherent heterogeneity of the agricultural systems existing across the EU, while still maintaining analytical consistency across the country models.

The AGMEMOD Partnership and its members aim to establish not only a coalition of economists working together across the EU, Accession States and EU neighbours, but also advisory circles of experts in commodity markets and agricultural sectors in each country, to review the models and projections. This process has led to the development of a core competency in the economic modelling of agricultural commodity markets and agricultural policy analysis, enhancing the quality of information available for policy and decision making at all levels.

High Level Motivation for Project

While policy reform remains a political process, policy makers increasingly use evidence based decision making in policy negotiations. Within the EU, Member States are free to adopt differing positions in respect of policy proposals, based on their assessment of the merits of the policy for their agriculture sector and wider economic and social interests. Those charged with developing policy proposals at

EU level, need to have an appreciation for the likely impact of a particular policy in order to identify, at an early stage, any issues that would prevent a policy proposal's acceptance by the Member States. In this context, a model such as the AGMEMOD model, which can provide Member State level detail, will be highly useful for EU and Member State based policy makers.

Motivation for Book

The motivation for writing this book is to provide fellow economists, policy analysts and other academics with a guide on how to build and operate a policy model of this kind and to help explain to policy makers the strengths and weaknesses of such models and the challenges which practitioners face in assessing the impact of policy change using this or similar models. A further objective of the book is to educate policy makers in how they should interpret the results of policy models.

How to Read this Book

The book is structured so that it can be read either in its entirety or by selected chapters. Academics and students may be interested in reading all chapters of the books, while policy makers may prefer to skip the more technical material on modelling (Chaps. 2 and 3).

Overview of Book

Chapter 1 provides a background to the model's development. It sets out the objective of the Common Agricultural Policy (CAP) and provides a brief history of the reforms of the CAP which have taken place over time, with a specific focus on the more recent reforms of the last 20 years. The heterogeneity of agriculture and agricultural policy across the EU Member States and the political implications which result in CAP negotiations are then discussed. Other modelling frameworks that have been developed in the past are then described. The justification for the modelling choices made in the design of the AGMEMOD model is then provided. Initially, some important issues associated with the interpretation of the model's results are discussed, including the definition of a baseline and the important distinction between projections and predictions/forecasts.

Chapter 2 describes the AGMEMOD model's structure. It provides a general description of the AGMEMOD model's structure including its country and commodity coverage. We present the general form of the model with specific examples of crops, livestock and dairy. Important features associated with the treatment of policy within the model are highlighted, in particular, market price support, direct payments and supply control. We describe how policy harmonisation addresses the incorporation of the diverse range of direct income supports. We discuss the concept of key price and key price equations that are used to link country models within the AGMEMOD model together and to close the AGMEMOD model at the EU27 level. We also explain how border protection and export competition measures are introduced in the model.

Chapter 3 describes the process of building, maintaining and using the AGMEMOD model. It describes the data collection, the database building, the conditions that data have to satisfy and how to adjust the data to ensure that commodity markets are balanced. The various types of policy data used in the model are described. Exogenous data such as macroeconomic indicators and world commodity prices are detailed and their sources are identified. Moreover, the chapter describes both the software used for data management, model estimation and the presentation of results as well as providing key information concerning the AGMEMOD software and user interfaces.

Chapter 4 presents the EU baseline outlook as generated by the AGMEMOD model. This baseline provides an example of the type of output produced by the model. The dissemination of analytic results is crucial to the wide acceptance of the quality of the results and their use in policy discussions. It is suggested that the format used in this chapter is a useful template in this regard. Results are provided for crops, livestock and dairy.

Chapter 5 analyses the impact of possible CAP policy changes using the AGMEMOD model. The main purpose of the chapter is to demonstrate a policy analysis application of the AGMEMOD model and to outline the types of policy options that can be explored using the AGMEMOD model. Specific results are provided for a scenario which examines the impact of equalising the level of decoupled direct payments per hectare across the EU.

Chapter 6 is the concluding chapter of the book. It draws together the key messages from the earlier chapters and explores the future capacity of the model in terms of possible commodity, country or other extensions to the AGMEMOD modelling framework. The potential usefulness for the modelling approach beyond the EU is also considered.

Websites

It is difficult to have a self-contained manuscript given the large number of commodities and national country markets considered. The reader is thus invited to visit the AGMEMOD website (http://www.agmemod.eu/) as well as the IPTS website (http://ipts.jrc.ec.europa.eu/publications/) to have access to the different studies conducted by the AGMEMOD Partnership. Furthermore, in order to obtain more completed information sets, the reader is invited to visit Springer's Extra Materials website (http://extras.springer.com/). The specific space dedicated to this book contains a demo version of the AGMEMOD model that will be as described in the next chapters

F. Chantreuil
K.F. Hanrahan
M. van Leeuwen
Editors

Chapter 1
Background

Kevin F. Hanrahan, Trevor Donnellan, and Emil Erjavec

Abstract The purpose of this chapter is to set the modelling and policy context for the AGMEMOD model. These contexts are important since they define the shape of AGMEMOD and the analytical purposes that it seeks to fulfil. The AGMEMOD model seeks to reflect the heterogeneity of European agriculture through its modelling of agricultural commodity markets in all EU Member States. The need for a detailed representation of policy instruments within the model's structure is also stressed given the heterogeneity in CAP implementation that has emerged since the Fischler reforms of 2003 and the accession of countries from Central and Eastern Europe. The strengths and weaknesses of partial equilibrium and general equilibrium approaches to agricultural policy modelling are also reviewed.

Keywords Origins of the CAP • CAP reforms • Modelling the CAP • Partial equilibrium modelling • Computable general equilibrium modelling

The purpose of this chapter is to set the modelling and policy context for the AGMEMOD model. This context is important since it defines to a large degree the shape of AGMEMOD and the purposes that it seeks to fulfil. Within the chapter we look at the history of the EU and the Common Agricultural Policy (CAP) and how this policy has evolved over time. We examine the modelling initiatives which have been developed to undertake policy analysis in agriculture and contrast them with the AGMEMOD framework. In so doing we aim to highlight the theoretical considerations that need to be reflected in economic models of EU agriculture.

K.F. Hanrahan (✉) • T. Donnellan
Rural Economy Research Centre, Teagasc, Athenry, Co Galway, Ireland
e-mail: kevin.hanrahan@teagasc.ie

E. Erjavec
Biotechnical Faculty, University of Ljubljana, Groblje 3, 1230 Domžale, Slovenia

F. Chantreuil et al. (eds.), *The Future of EU Agricultural Markets by AGMEMOD*,
DOI 10.1007/978-94-007-2291-0_1, © Springer Science+Business Media B.V. 2012

Economic models of the agricultural sectors of the EU and other regions need to both account for the past but also to look forward to the future in terms of their design by incorporating mechanisms that allow for the analysis of future policy changes. Model practitioners need an awareness of historical policy developments and political debates surrounding current and future policy reforms. This knowledge aids in the understanding of past developments in the sector and also allows for an educated assessment of the likely future direction of policy. In anticipating possible future policy paths awareness of the drivers of change both from within and from outside the EU is required (Josling 2008).

1.1 Origins of EU and CAP

Following World War II, co-operation among the countries of Europe was seen as a key step in preventing wars in the future. Inspired by the French Foreign Minister, Robert Schuman, the organisation which we today recognise as the European Union, began with six member countries of the European Coal and Steel Community in the 1950s. In 1957, this co-operation was extended through the signing of the Treaty of Rome, that created the European Economic Community.

Agricultural policy was an important element of the Treaty of Rome, with articles 32–38 of the Treaty setting the objectives and parameters for the Community's agricultural policy. The CAP and its constituent commodity organisations were developed over the following decade. The original objectives for European agricultural policy within the Treaty of Rome were:

- to increase agricultural productivity by promoting technical progress and by ensuring the rational development of agricultural production and the optimum utilisation of the factors of production, in particular labour;
- thus to ensure a fair standard of living for the agricultural community, in particular by increasing the individual earnings of persons engaged in agriculture;
- to stabilise markets;
- to assure the availability of supplies;
- to ensure that supplies reach consumers at reasonable prices.

The Gothenburg Agenda (EU 2001) added sustainable development as an objective, while the Lisbon Treaty (EU 2008) added consumer protection and animal welfare objectives. Since the Agenda 2000 (EC 1999) agreement objectives relating to rural development have received increasing emphasis, but these are still not reflected in the Treaty's objectives for agriculture policy (Bureau and Mahé 2008).

The CAP which was developed under the Treaty of Rome is based on the three principles:

- market unity;
- community preference; and
- financial solidarity.

Market unity reflects the single market basis of the European Union. Community preference reflects the protection of agriculture behind a common external tariff. Financial solidarity (perhaps increasingly the most important principle) is reflected in the extent to which the CAP is financed by a central EU budget as opposed to being co-financed by the Member States. Currently all market support and direct income support payments (so-called Pillar I expenditures) are funded by the EU budget (the European Agricultural Guarantee Fund) while co-financing is the norm for the different "axes" of Rural Development policy (so-called Pillar II of the CAP).

1.1.1 Recent CAP Reforms

While expenditure linked to the CAP represents a declining share of the overall EU budget, it remains one of the most important of EU *common* policies. Over the period 2007–2013 CAP spending will account for close to 40% of total EU spending annually (Gros 2008).

Since its creation in 1957 the CAP has, through a series of reforms, evolved towards the system that exists today. The early CAP was characterised by price supports designed to improve farmer incomes and ensure sufficient food supplies. These price supports were "successful" in that they led to considerable food surpluses in particular sectors. These surpluses were exported with the aid of subsidies where possible or were otherwise stored as intervention stocks. The CAP attracted criticism as a result due to the considerable cost of these measures and due to their impact on world agricultural commodity markets. By the mid-1980s supply controls began to be introduced in some sectors to limit the cost of dealing with the CAP engendered agricultural commodity surpluses and in subsequent years intervention stock holding became more limited as the level of price support was reduced (Josling 2008).

By the late 1980s there was a growing consensus among policy makers that the CAP would need to become more market focused. Over the period from 1990 to 2010 the CAP has been reformed on several occasions. Table 1.1 summarises these successive reforms.

The MacSharry reforms, agreed in 1992, heralded the beginning of a reduction in the price support provided to some commodities (cereals and beef predominantly) from 1994 onwards (Swinbank and Tanner 1996). Under the MacSharry reforms, the expected loss in income from the market place, as a result of the lowering of guaranteed support prices, was compensated by an increase in the level of coupled direct income support to producers. The motivations for this reform were both internal and external, with internal political pressures to control budgetary spending on agriculture and the desire on the part of the EU to conclude the Uruguay Round of the General Agreement on Tariffs and Trade.

The CAP reform process was continued in the Agenda 2000 reforms that were agreed in 1999. Cereal and beef support prices were further reduced and a commitment was made to reduce dairy price support by the middle of the following decade,

Table 1.1 Recent CAP reforms

Reforms	Year of agreement	Main motivation
MacSharry CAP reforms	1992	Internal political pressures to control budgetary spending on agriculture and negotiations on agriculture with the General Agreement on Tariffs and Trade (GATT) Uruguay Round
Agenda 2000	1999	To control the budgetary cost of the accession of Central and Eastern European countries to the EU
Fischler reforms (MTR)	2003	Increase the market focus of EU agriculture and strengthen the position of the EU within the WTO Doha Development Agenda (DDA) negotiations
CAP health check	2008	To increase the market focus of the dairy sector in the EU and to further reduce the cost of agricultural support

with the expected negative impact on producer prices of lower intervention prices to be compensated through the provision of further direct income support. These reforms were motivated by the desire to control the budgetary cost of the anticipated accession of the countries of Central and Eastern Europe (the so called new Member States) following the collapse of the Soviet Union.

A Mid Term Review of the CAP, known as the Fischler reforms, was agreed in 2003 and resulted in further changes to the CAP. The main focus of this reform was the introduction of decoupled direct income support payments, breaking the link between the provision of support and the requirement to produce output. Increased spending on rural development policy was also agreed. The main motivations of the Fischler reforms were to increase the market focus of EU agriculture, to better align the CAP for future WTO negotiations, specifically to reduce the extent to which the CAP could be viewed as trade distorting, and to accommodate the cost of EU expansion within the EU agriculture budget.

Most recently in 2008, agreement was reached on what was termed the CAP Health Check. The key motivation of the CAP Health Check was to increase the market focus of the dairy sector in the EU and to further reduce the cost of agricultural support within the overall EU budget. Under the reform it was agreed that EU milk quotas (supply controls) would be abolished in 2015, and that this abolition would be preceded by a series of increases in EU milk quota. In the reform, spending on rural development policy was increased with funds for this increase in spending coming from an increase in the rate of modulation applied to the first Pillar direct income support payments EU farmers receive. Modulation was introduced to the CAP in the Fischler reforms of 2003 and is a deduction of a percentage of the payment entitlement of farmer, where payments exceed a set threshold. The funds raised through modulation are used to support rural development and other agricultural policy objectives.

1.1.2 Future Evolution of CAP

Internal EU drivers for reform will remain important factors in the future evolution of the CAP. The current CAP is largely based on its direct payment systems which account for 70% of total CAP expenditure and over 89% of first Pillar spending. Most of this spending is on decoupled income support payments under the Single Payment System (SPS) that operates in EU15 Member States, and the Single Area Payments Scheme (SAPS) that operates in most of the new Member States. Changing the shape of the CAP will necessarily involve changing the direct payment systems and any changes will affect incomes in the agriculture sector and the net flow of funds from the CAP to the EU Member States. Any changes to the financial solidarity principle of the CAP would have additional implications for the allocation of funds between Member States, with any further erosion of the financial solidarity principle implying reductions in the magnitude of the benefits that accrue to current net beneficiaries and some concomitant reduction in the negative operating balances of current net contributors.

There are also factors outside the EU, which will increasingly play a role in the future shape of the CAP. To some degree, the future shape of the CAP will be determined by external, as well as internal, constraints, including the international trade agreements negotiated by WTO members. Current agricultural trade rules are based on the Uruguay Round Agreement on Agriculture and any successor agreement that may arise on the conclusion of the Doha Development Agenda round, will limit the extent to which the EU could return to the agricultural policy framework of the 1970s and 1980s.

The ongoing WTO negotiations on agriculture are focused on what are termed the three pillars of *trade distorting domestic support*, *export competition* and *markets access* (WTO 2008). Despite the collapse of negotiations in 2008, WTO meetings in Geneva have continued. Although the timeline for a conclusion to the Doha Round has drifted, it seems inevitable that any agreement that might be reached will further liberalise agricultural trade. Any future agreement will further constrain the freedom of WTO Members to link direct payments to agricultural production, limit the degree to which WTO members can protect their agricultural goods markets from competition from other WTO members and also limit the degree to which WTO members can subsidise of otherwise support the export of agricultural goods.

The implication of these characteristics of the future EU external environment is that EU agricultural commodity markets will become increasingly exposed to international (i.e. extra-EU) competition. The tariff based wedge between EU and non-EU agricultural commodity markets will diminish over time allowing the level of imports into the EU to increase.

1.2 Modelling the CAP at Member State Level

This section provides the justification for modelling the EU agriculture at the Member State level noting the degree of heterogeneity that exists across EU agriculture.

1.2.1 Heterogeneity of EU Agriculture

EU agriculture is very diverse. The agriculture sectors of the different Members States have very different specialisations (based on differences in soil and agro-climatic conditions) and also very different farm structures.

In Fig. 1.1 the proportion of agricultural output accounted for by three categories of agricultural commodities are shown. The commodity groups are (1) ruminant meat and dairy, (2) cereals, oilseeds, sugar beet and potatoes, and (3) fruits, vegetables, wine and olive oil. Dramatic differences between Member States are apparent. In Ireland, the United Kingdom, Finland and Luxembourg the agricultural sector output is dominated by ruminant meat and dairy production. In contrast, in Member States such as Greece, Italy, Spain and Portugal fruits, vegetables, wine

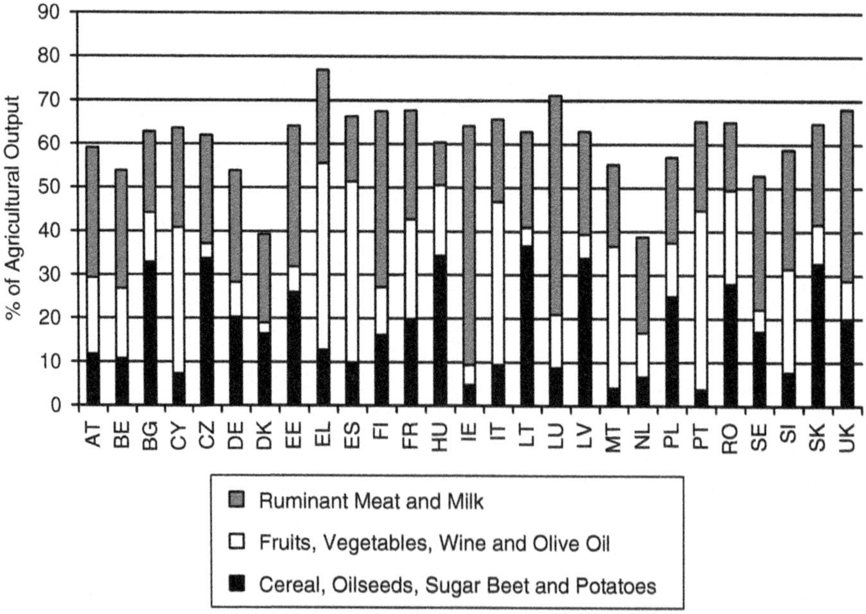

Fig. 1.1 Share of products in agricultural production (Source: Agriculture in the European Union – Statistical and Economic Information 2010, Commission of the European Communities (2011))

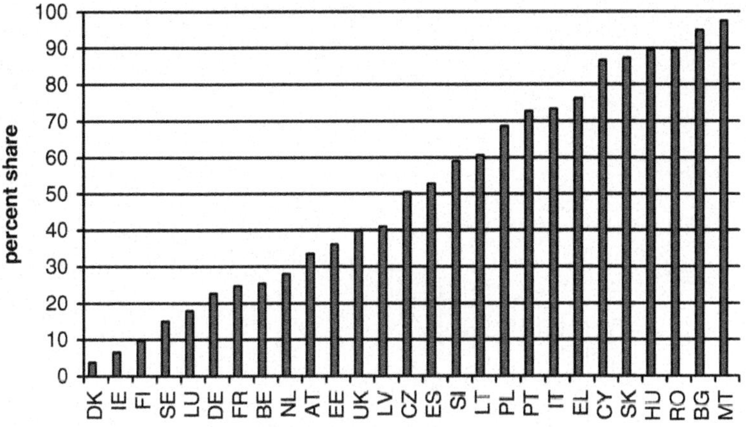

Fig. 1.2 Proportion of farm population of less than 5 ha (UAA) (Source: Agriculture in the European Union – Statistical and Economic Information 2010, Commission of the European Communities (2011))

and olive oil dominate agricultural output. Farm structures are also very different across the EU. In Fig. 1.2 the proportion of total farm numbers in each Member State of less than 5 ha of utilised agricultural area (UAA) is shown. In some Member States agricultural production is dominated by small farm holdings while in other larger holding are more prevalent.

These two aspects of European agriculture provide a partial rather than a comprehensive picture of the underling heterogeneity of agriculture in the EU, nevertheless they indicate that a modelling framework with the capacity to reflect the underlying heterogeneity of EU agricultural production may be preferable to one in which such Member State differences are not captured. By adopting a country by country modelling approach, the AGMEMOD model allows the individual country models, through their parameterization and commodity coverage, to better reflect the diversity in agro-climactic conditions, agricultural structures and agricultural output that exists between EU Member States' agricultural sectors. Through this bottom up approach the AGMEMOD model aims to better capture the heterogeneity of EU agriculture.

1.2.2 *Heterogeneity of EU Agricultural Policy*

The CAP as currently structured is largely based on protection of the EU market through the imposition of import tariffs and the payment of direct income supports to farmers. While the market price support and direct income support elements of the CAP are financed from a common budget, recent reforms and the expansion of the EU in 2004 and 2007 introduced considerable heterogeneity to the *common* agricultural policy.

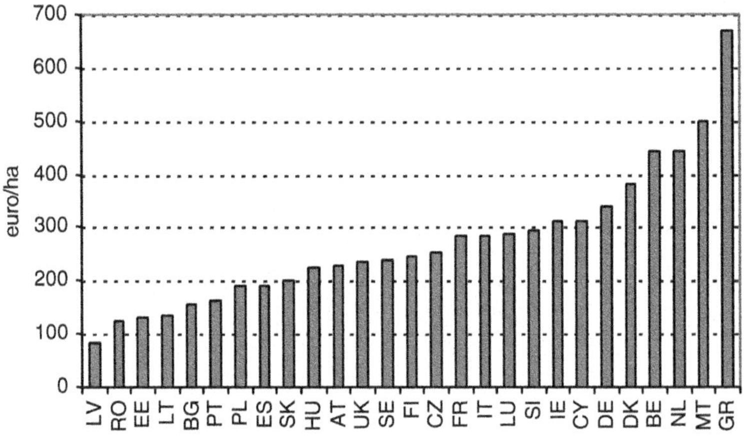

Fig. 1.3 Direct income support payments in EU Member States (Source: Commission of the European Communities 2009)

In the so-called old Member States (EU members prior to 2004), direct income supports are now mostly paid in the form of decoupled direct payments. Among EU15 Member States, a number of different models for determining the level of direct income support per hectare are used. These range from models where the coupled payments received by a given farmer in the years 2000–2002 are paid to that farmer (the so-called historical model), to systems where the sum of all payments received in a Member State (or region of a Member State) are divided evenly across all hectares of eligible agricultural land (the so-called flat area payment model). Most direct income support payments are decoupled from production though some coupled payments have been retained by some Member States. Following the 2008 CAP Health Check, the only coupled direct payments that may still be paid are to farmers of beef cows (suckler cow premium) and sheep (ewe premium).

In most of the Member States that joined the EU in 2004 and 2007, a flat area payment scheme, the SAPS operates. Under the terms of the accession of these countries to the EU, it was agreed that the CAP would be phased in over the period to 2013. Over this CAP phasing in period, the EU12 Member States have the freedom to top up the EU budget funded direct income support payments from their national exchequer via complementary national direct payments (CNDP). Despite this freedom, the level of budgetary support (both national and EU) to farm incomes in the newer EU12 Member States, is generally lower per hectare than in EU15 Member States. Figure 1.3 illustrates the large differences in direct income support payments across EU Member States, with support per hectare ranging from €79 per hectare in Latvia, to €681 per hectare in Greece. The large disparities in the direct income support per hectare across the EU are reflected in large differences in aggregate national CAP receipts across EU Member States. These disparities have been a source of considerable controversy in the EU (Begg 2005).

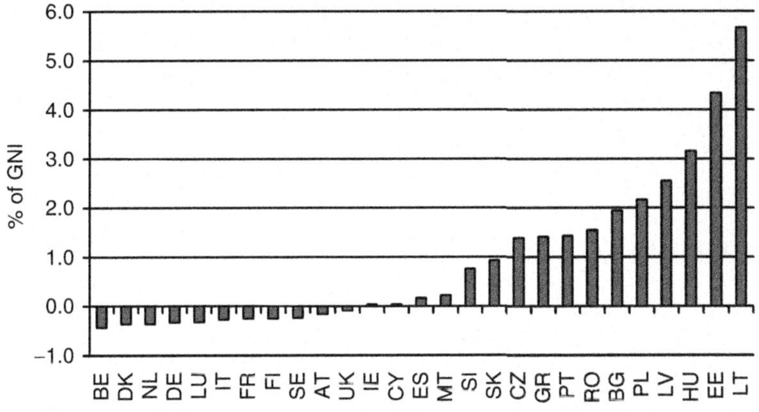

Fig. 1.4 2009 Operating budgetary balance in EU Member States (Source: Commission of the European Communities 2010)

1.2.3 Heterogeneity of CAP Outcomes

A notable feature of the CAP reform discussions at EU level is that Member States frequently adopt conflicting positions. Richter (2008) has commented on the "juste retour" attitude, where an objective of each EU Member State during the policy reform negotiations is to secure the best possible net financial position vis-à-vis the EU budget rather than the achievement of some other objective(s) immediately related to the proposed policy reform. Under this approach, the impacts of policy reforms in agriculture are assessed by negotiators, not only on the basis of their impact on agricultural production and agricultural incomes, but also on their impact on the budgetary position of each Member State. Within the EU budget there are distinct groups of Member States that are net payers and net beneficiaries, the pattern of "winners" and "losers" is affected importantly by the CAP, and as a consequence the political success of agricultural policy reform proposals is affected by their impact on Member State net balances. Figure 1.4 presents the operating budgetary balances of the EU Member States as a percentage of Gross National Income (GNI). Agricultural policy reforms that dramatically alter the pattern of budgetary flows between Member States are unlikely to succeed given the strong status quo bias of the EU budgetary process (Gros 2008). The EU Budget Review and the parallel negotiations on the shape of the CAP post 2013 provide an opportunity, perhaps, to remove the link between European agricultural policy and EU budgetary controversies (Bureau and Mahé 2008).

Given the heterogeneity of agricultural structures and the diversity in agricultural policy as implemented across Member States, the modelling of agricultural policy at the aggregate EU27 level may not capture important differences in both policy implementation and outcome across Member States. These differences in policy implementation and impact are of intrinsic interests to policy makers in the different Member States. Given the political process through which EU agricultural policy is formed (where Member States and the European Parliament have the role of deciding on any policy changes) these differences are also of interest to policy makers in the EU's institutions based in Brussels.

1.3 Model Features

Due to the heterogeneity of European agriculture and the heterogeneity of European agricultural policy there is a need for models to address the diversity of:

- EU agriculture at the Member State level;
- CAP and its implementation at the Member State level;
- results of policy options and decisions at the Member State level.

These three considerations point to the need for an economic model of EU agriculture, which is sufficiently disaggregated to examine agriculture at the Member State level and which ideally involves practitioners from the Member States in that modelling activity with specialist knowledge of agriculture in their country. Such a model of EU agriculture would be distinct from most that currently exists which have tended to operate at an EU level of aggregation.

The central role of modellers, based within each EU Member State, has also facilitated the development of an interactive review process of the AGMEMOD model's results. This process involves industry and government policy makers and is valuable in improving the usefulness of the analysis. Potential disadvantages of the bottom up approach include the need for careful coordination of model development and database maintenance activities across a very large set of partners (Salamon et al. 2008).

1.4 Economic Models of European Agriculture

A number of different tools such as computable general equilibrium (CGE) models and static or dynamic partial multi-market equilibrium models have been used to analyse the likely impact on agricultural markets of changes in Member States' domestic agricultural policies, the CAP, and liberalisation of agricultural trade through future WTO or other bilateral trade agreements.

Jensen (1996) distinguished the following modelling categories that are commonly used in the area of agricultural policy analysis:

- international trade models;
- national economy-wide models;

- sectoral models;
- partial equilibrium market models for individual or a group of agricultural commodities; and
- individual farm level models.

In modelling agricultural policy reforms, the most common modelling approaches used tend to be CGE or partial equilibrium (PE) models.

Models based on the CGE theory are designed to represent the overall functioning of a national economy. Formulation of the interactions between sectors of an economy in CGE models is flexible in aggregation and defined using input-output tables or social accounting matrices that characterise money flows between different sectors in the economy. Generally speaking factor of production are mobile between sectors of the economy and the equilibrium prices of all products are determined simultaneously (Robinson 1989; Pindyck and Rubinfeld 2008).

However, in CGE models agricultural production is often aggregated (sometimes even within a single account) so as to limit the complexity of the model and to improve its computational feasibility. In addition, inclusion of some agricultural policy measures is often difficult due to aggregation of agricultural production and inadequate representation of physical resource constraints (Banse and Tangermann 1996). Tyers and Anderson (1992) note that due to such aggregation choices, the interaction and causal linkages between different agricultural production sectors are often weak in CGE models.

On the other hand, and by definition, PE models do not include linkages that allow for the analysis of the impact of developments in the agricultural sector on other sectors of the economy. However, as PE models have the ability to incorporate greater amounts of detail on production and policy instruments, they have advantages over their CGE counterparts (Salvatici et al. 2001). PE models generally describe one sector or a group of closely related products in an economy with a greater level of disaggregation than is common in CGE models. Due to the capacity of PE models to incorporate detailed representations of relationships between policy instruments and agricultural commodity supply and demand, these types of models are very suitable to the analysis of the agricultural sector of developed economies. The PE framework also facilitates the extensive coverage of agricultural commodities and countries. Important features of the PE models are their relatively simple economic structure, and their relatively easily understandable and interpretable results. This last feature can be advantageous when model results are used by non-economists. A more detailed overview of both general and partial equilibrium models used in agricultural policy analysis and their different features can be found in Van Tongeren et al. (2001).

A drawback of many existing PE and CGE models is the often limited set of policy instruments (particularly regarding the CAP) that are incorporated in the model structures and databases. This policy representation issue, as well as the need to capture the heterogeneity of the agricultural sector at the EU Member State level, were the principal motivations for developing the AGMEMOD model.

1.5 The Baseline Concept

In Chap. 4 we describe how the AGMEMOD model is used to produce a 10 year forward baseline projection for the main agricultural commodities in the EU. Although the baseline represents a projection of commodity prices, production and quantities traded, readers should note that the generation of these projections is not the main objective of the AGMEMOD model or modelling process. The main purpose of AGMEMOD is the analysis of the impact of policy measures, either proposed or actual, and the quantitative assessment of the effects of policy change. The baseline, no policy change, projections allow us to highlight key medium term market developments and draw some conclusions about future policy developments and their likely impact on EU agriculture. It is important to understand that the baseline projections in this book are not 'forecasts' or 'predictions', but are projections based on (1) a well-defined set of assumptions and (2) a set of models of European agricultural commodity markets. The primary function of the baseline is as a counter-factual state of the world that allows the analyst to assess the impact of a given policy change.

A characteristic of baseline projections is that, when represented graphically, they tend to follow smooth paths and can often contrast with the more jagged patterns observed in historical data. It is important to understand that this historical variation is often the product of unforeseen supply or demand shocks in the market under consideration. On the supply side these shock can be due to abnormally favourable or unfavourable weather conditions, wars, or animal or plant disease outbreaks all of which can impact on production and trade in a particular period. On the demand side shocks also give rise to variability but this variability is usually less pronounced than is the case with shocks to the supply side. Demand shocks can arise due to, among other things, human health concerns (BSE), such as sudden shifts in dietary preferences (Atkins diet), economic shocks (recession) or the development of new uses for commodities (emergence of biofuel production). In general, economic models of agriculture cannot anticipate these types of supply or demand shocks. As a consequence baseline projections tend to be smooth and should be seen as an indicator of the likely evolution of production, consumption and prices in a particular market under a specific set of circumstances rather than a definitive forecast of the future.

Aside from the unanticipated shocks detailed above, there are other reasons to expect that a baseline projection is unlikely to materialise. Policy changes, due to CAP reform or changes resulting from WTO Doha Development Agenda will emerge and these policy changes will impact on production decisions. Chapter 5 presents the results of a CAP reform scenario that has been examined using the AGMEMOD model. The analysis reported illustrates how an alternative scenario to the baseline can be developed and how the difference between that scenario outcome and the baseline can be used to assess the economic impact of a proposed policy change.

The generation of baseline projections is complicated by the need to use up to date market data. EU official data (from Eurostat or a Member State's statistical office) often takes time to compile, verify and disseminate. This can mean that the most recent observations for particular data can be 1 or 2 years old. As a result at any point in time a data "gap" between the past and the present is present that needs to be filled from unofficial or industry data sources. In Chap. 3 this data capture and management issue is discussed.

1.6 Conclusion

Two general drawbacks of existing models of the European agricultural market that are used for policy analysis are identified. The first is the limited set of CAP instruments that are incorporated in their model structures. The second drawback of existing models is that the modelling of European agricultural markets takes place at an aggregate EU or supra-Member State level. The need to represent the heterogeneity of the agricultural sector at the detailed EU member State level and to have a complete and comprehensive incorporation of CAP instruments was the principal motivation behind the development of the AGMEMOD model.

The AGMEMOD model is a dynamic partial equilibrium (PE), multi-commodity, multi-country model designed specifically to cater for the needs of detailed EU agricultural policy analysis (Riordan et al. 2002). Through a bottom up approach to model development (detailed in the following chapter), the model reflects the heterogeneity of European agriculture and of CAP policy as implemented in the 27 EU Member States. In the following chapters the structure of the model, its operation and examples of the model's use are presented.

References

Banse M, Tangermann S (1996) Agricultural implications of Hungary's accession to the EU. Partial vs. general equilibrium effects. Working Paper 1/2. Joint Research Project "Agricultural Implications of CEEC Accession to the EU", Göttingen

Begg I (2005) Funding the European Union. A federal trust report on the European Union's Budget. The Federal Trust for Education and Research, London

Bureau JC, Mahé LP (2008) CAP reform beyond 2013: an idea for a longer view. Notre Europe Studies and Research No. 64, Notre Europe, Paris

European Commission (2010) EU budget 2009: financial report. Publications Office of the European Union, Luxembourg

European Commission (2011) Agriculture in the EU: statistical and economic information report 2010. Publications Office of the European Union, Luxembourg

European Council (1999) Council regulation (EC) No 1257/1999 of May 17, 1999 on support for rural development from the European Agricultural Guidance and Guarantee fund (EAGGF) and amending and repealing certain Regulations. OJ L160, 26 June, 1999

European Union (2001) Göteborg European Council 15 and 16 June 2001: presidency conclusions. SN 200/1/01 REV 1. Available via http://www.consilium.europa.eu/uedocs/cms_data/docs/pressdata/en/ec/00200-r1.en1.pdf. Cited 21 June 2011

European Union (2008) Consolidated versions of the treaty on European Union and the treaty on the functioning of the European Union. Office for Official publications of the European Communities, Luxembourg, OJ C 115, 9.5.2008

Gros D (2008) How to achieve a better budget for the European Union? CEPS working document No. 289, April 2008. Centre for European Policy Studies (CEPS), Brussels

Jensen JD (1996) An applied econometric model for Danish Agriculture (ESMERALDA), Danish Institute of Agricultural and Fisheries Economics, report no. 90

Josling T (2008) External influences on CAP reforms: an historical perspective. In: Swinnen JFM (ed) The perfect storm: the political economy of the Fischler Reforms of the Common Agricultural Policy. Centre for European Policy Studies (CEPS), Brussels

Pindyck R, Rubinfeld D (2008) Microeconomics, 7th edn. Prentice Hall, Upper Saddle River

Richter S (2008) Facing the monster "Juste retour": on the net financial position of member states vis-à-vis the EU Budget and a proposal for reform. Research Report 348, Institute for International Economic Studies, wiiw, Vienna

Riordan B, Donnellan T, Hanrahan KF, McQuinn K (2002) Projection of policy impacts on the agri-food sector: overview and introduction to AG-MEMOD. Paper presented at the 10th Congress of the European Association of Agricultural Economists (EAAE), Zaragoza, 28–31 August 2002

Robinson S (1989) Multisectoral models. In: Chenery HB, Srinivasan TN (eds) Handbook of development economics. North Holland, Amsterdam

Salamon P, Chantreuil F, Donnellan T, Erjavec E, Esposti R, Hanrahan KF, van Leeuwen M, Bouma F, Dol W, Salputra G (2008) How to deal with the challenges of linking a large number of individual national models: the case of the AGMEMOD Partnership. Agrarwirtschaft 57(8):373–378

Salvatici L, Anania G, Arfini F, Conforti P, De Muro P, Londero P, Sckokai P (2001) Recent developments in modelling the CAP: hype or hope? In: Heckelei T, Witzke HP, Henrichsmeyer W (eds) Agricultural sector modelling and policy information systems. Proceedings of the 65th European Seminar of the European Association of Agricultural Economists (EAAE). Bonn, Germany, 29–31 March 2000. Wissenschaftsverlag Vauk, Kiel

Swinbank A, Tanner C (1996) Farm policy and trade conflict: the Uruguay Round and CAP reform. University of Michigan Press, Ann Arbor

Tyers R, Anderson K (1992) Disarray in world food markets: a quantitative assessment. Cambridge University Press, Cambridge

Van Tongeren FW, Meijl H, Surry Y (2001) Global models applied to agricultural and trade policies: a review and assessment. Agr Econ 26(2):149–172

World Trade Organization (WTO) (2008) Revised draft modalities for agriculture. WTO, TN/AG/W/4/Rev.4, Committee on Agriculture Special Session, Geneva

Chapter 2
Model Structure and Parameterisation

Roberto Esposti, Guna Salputra, Frédéric Chantreuil, Kevin F. Hanrahan, Petra Salamon, and Andrzej Tabeau

Abstract This chapter provides a general description of the AGMEMOD model's structure including its country and commodity coverage. We present the general form of a PE model and the AGMEMOD general form with specific examples of crops, livestock and dairy. Important features associated with the treatment of policy within the model are highlighted, in particular, market price support, direct payments and supply control. We describe how policy harmonisation addresses the incorporation of the diverse range of direct income supports and discuss the concept of key price and key price equations. The introduction of border protection and export competition measures in the model is also discussed.

Keywords Partial equilibrium model • AGMEMOD general form • Model specification • Model estimation • Model closure • Price formation • Policy harmonisation

R. Esposti (✉)
Department of Economics, Università Politecnica delle Marche, Ancona, Italy
e-mail: r.esposti@univpm.it

G. Salputra (✉)
Latvia State Institute of Agrarian Economics, 14 Struktoru St, Riga, Latvia
e-mail: guna@lvaei.lv

F. Chantreuil
Université de Caen Basse Normandie, UFR Sciences Economiques et Gestion,
17 rue Claude Bloch, BP 5186, F-14032 Caen Cedex, France

K.F. Hanrahan
Rural Economy Research Centre, Teagasc, Athenry, Co Galway, Ireland

P. Salamon
Johann Heinrich von Thünen-Institute (vTI), Brunswick, Germany

A. Tabeau
LEI Wageningen UR, The Hague, The Netherlands

F. Chantreuil et al. (eds.), *The Future of EU Agricultural Markets by AGMEMOD*,
DOI 10.1007/978-94-007-2291-0_2, © Springer Science+Business Media B.V. 2012

In this chapter, we provide a general description of the AGMEMOD model's structure including its country and commodity coverage. We present briefly the general form of partial equilibrium (PE) models and the AGMEMOD general form in detail, emphasising to what extent the AGMEMOD modelling framework differs from PE general form. Important features associated with the treatment of policy within the AGMEMOD model are highlighted, in particular, market price support, direct payments and supply control. The discussion of these policy instruments is linked to the discussion of CAP developments in Chap. 1. The need to adapt or modify economic models for agriculture and policy changes is emphasised. The approach adopted, termed policy harmonisation, allows for the incorporation of a diverse range of different direct income supports in a consistent fashion. The concept of the key price that is used in the AGMEMOD model is discussed and a description of the key price equations used in the AGMEMOD model is provided. We also explain how border protection and export competition measures are introduced in the AGMEMOD model.

2.1 PE Model Structure

In a PE model of the agricultural commodity markets for a given country, each commodity sub-model contains the behavioural response of economic agents (farmers, consumers, etc.) to changes in market prices, policy instruments, other exogenous variables (such as world prices) as well as to the lagged endogenous variables that determine the model's recursive structure. For each commodity modelled the set of behavioural equations must include equations that determine the supply (beginning stocks, production and imports) and the demand sides (domestic use, exports and ending stocks) of the market. These supply and demand equations define how, in any given year, equilibrium (i.e., supply equals demand) is found within the single commodity market. The dynamic behaviour of the model is determined by the model's recursive structure, i.e. through lagged endogenous variables entering as exogenous determinants of the current period's equilibrium supply and/or demand. For each commodity market model a supply side and demand side identity must be chosen to close the model and ensure that for all time periods, supply equals demand. Thus, for each time period equilibrium is ensured at the market clearing commodity price.

PE commodity models can be aggregated with similar models representing other commodity markets to form a multi-commodity PE model of the agricultural sector for a given country. Such country-level PE agricultural commodity models can themselves also be further aggregated to form multi-country multi-commodity models. At each level of aggregation, variables that at a previous level were exogenous become endogenous and the complexity of the model increases. Figure 2.1 depicts the general form of a single PE commodity market model within the AGMEMOD modelling approach. In the next section, the details of this modelling approach (a non-spatial, multi-country, multi-commodity PE model) are presented.

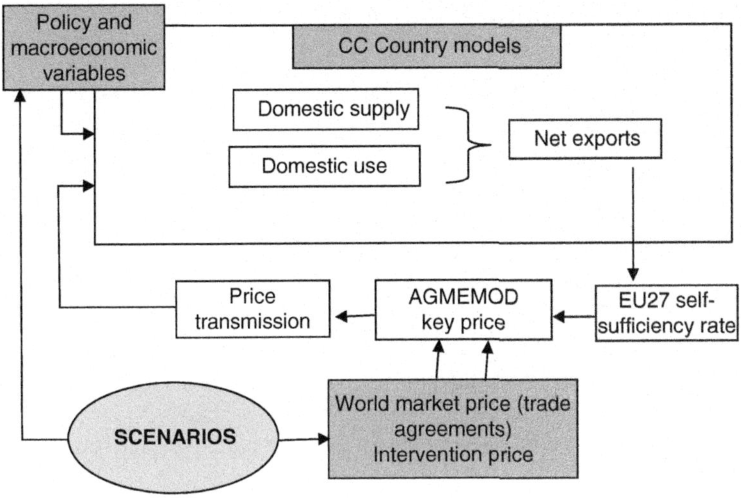

Fig. 2.1 General structure of an AGMEMOD country level agricultural commodity model

2.2 AGMEMOD Model Structure

2.2.1 Country and Commodity Coverage

The modelling "space" considered in the AGMEMOD framework is thus defined by the combination of two levels: countries and commodities. Along the country level dimension, AGMEMOD covers all EU27 countries plus individual countries of the Western Balkan Region (Croatia and Macedonia), the Black Sea Region (Ukraine and Russia) and Turkey. These non-EU country models, although more stylised and simplified, are constructed according to the same general rules and templates and, therefore, are designed to fully combine with the models developed for the EU countries. These non-EU country models are included either because they are EU Accession States (Croatia, Macedonia and Turkey) or because they are of major importance to EU commodity markets (e.g. Ukraine and Russia are very important to EU cereal markets). Along the commodity dimension, the AGMEMOD model includes the following commodity markets:

- soft wheat, durum wheat, barley, maize, rye, oats, triticale, rice;
- soybean, rape seed, sunflower seed, vegetable oils and meals;
- potatoes, sugar beets, sugar;
- beef and veal, pork, poultry, sheep and goats, eggs;
- milk, butter, skimmed milk powder, cheese, whole milk powder, cream, casein, other fresh dairy products;
- wine, cotton, tobacco, olive oil;
- apples, oranges and tomatoes.

2.2.2 Functional Form

The general structure of the AGMEMOD country and composite models is based on the template that can be found in Hanrahan (2001). A more detailed model description can be found in Chantreuil, Hanrahan and Levert (2005). The form of the model template varies across four different groups of commodities (1) grains, oilseeds and root crops, (2) permanent crops, (3) livestock and (4) dairy products.

For crops (cereals, oilseeds and root crops) the first equations on the supply-side concern agricultural land allocation. Land allocation in the AGMEMOD model is a two-step process. In the first step producers choose the total land area allocated to the aggregate of the grains, oilseeds and root crop culture groups (i). Then, in a second step, each crop j is allocated its own share of the total area of the corresponding crop culture group (i).

The equation of the total harvested area for grains, oilseeds and root crops can be written as

$$ah_{i,t} = f\left(p^j_{i,t-1}, ah_{l,t-1}, V\right), \quad j = 1,...,n; \quad i,l = 1,...,3; \quad i \neq l \tag{2.1}$$

where $ah_{i,t}$ and $ah_{l,t}$ are the areas harvested in year t for culture group i and for culture group l, respectively. $p^j_{i,t-1}$ is the real price in year $t-1$ of crop j belonging to the culture group i (all n crops or just the most relevant are considered), and V is a vector of exogenous variables which could have an impact on the area of culture group i harvested. Such variables include, *inter alia*, policy variables such as the rate of arable aid compensation and the set-aside rate. A detailed description of how policy support enters the supply-side of the AGMEMOD commodity models is provided in Sect. 2.4.

The equations used to determine the share of crop j belonging to as culture group i ($sh^j_{i,t}$) can be written as:

$$sh^j_{i,t} = f\left(p^k_{i,t-1}, sh^j_{i,t-1}\right), \quad j,k = 1,...,n. \tag{2.2}$$

Thus the jth crops share depends on its own price $p^j_{i,t-1}$ as well as on the price of any other kth crop of the same ith culture group ($p^k_{i,t-1}$). The yield equations of crop j in culture group i can be written as:

$$r^j_{i,t} = f\left(p^j_{i,t-1}, r^j_{i,t-1}, V\right), \quad j = 1,...,n \tag{2.3}$$

where $r^j_{i,t}$ is the yield per hectare of crop j belonging to culture group i, and V is a vector of exogenous variables that could have an impact on the yield of the crop being modelled.

The total production of crop j belonging to culture group i in year t ($PR^j_{i,t}$) is then determined as the product of area allocated to culture group i, the share of crop j and the yield per hectare of crop j, that is: $PR^j_{i,t} \equiv ah_{i,t} \cdot sh^j_{i,t} \cdot r^j_{i,t}$.

On the demand side, per capita non-food (crush and feed demand) and food (or non-feed) uses are modelled using the following general functional forms:

$$Fu_{i,t}^{j} = f\left(p_{i,t}^{k}, Z\right), \quad j,k = 1,...,n \tag{2.4}$$

where $Fu_{i,t}^{j}$ is the feed demand for the *j*th crop belonging to culture group i and Z a vector of endogenous variables, which could have an impact on the demand considered (e.g. meat production).

$$NFu_{i,t}^{j} = f\left(p_{i,t}^{k}, NFu_{i,t-1}^{j}\right), \quad j,k = 1,...,n \tag{2.5}$$

where $NFu_{i,t}^{j}$ is the food (non-feed) demand for *jth* crop belonging to culture group i. The total domestic use (or demand) of crop j belonging to culture group i in year t ($DU_{i,t}^{j}$) is thus calculated as $DU_{i,t}^{j} \equiv Fu_{i,t}^{j} + NFu_{i,t}^{j}$.

In modelling the supply and use of grains, oilseeds and root crops, the supply and use of oilseed oil and meals are modelled. The supply sides of the oil and meal markets associated with each oilseed commodity are determined by the quantity of oilseeds crushed and by technical coefficients relating quantity of oilseed crushed to quantity of oil and meal output. The crush demand for oilseed crop j ($CR_{i,t}^{j}$) is modelled as:

$$CR_{i,t}^{j} = f\left(p_{O,i,t-1}^{k}, p_{M,i,t-1}^{k}, CR_{i,t-1}^{j}\right), \quad j,k = 1,...,n \tag{2.6}$$

where $p_{O,i,t-1}^{k}$ is the price of the respective seed oil and $p_{M,i,t-1}^{k}$ the price of the respective seed meal produced from the crushing process. In such cases the total domestic use is calculated as $DU_{i,t}^{j} \equiv Fu_{i,t}^{j} + NFu_{i,t}^{j} + CR_{i,t}^{j}$.

The stock, export and import equations, in general, have the following functional forms:

$$ST_{i,t}^{j} = f\left(PR_{i,t}^{j}, DU_{i,t}^{j}, ST_{i,t-1}^{j}\right) \tag{2.7}$$

$$EX_{i,t}^{j} = f\left(PR_{i,t}^{j}, DU_{i,t}^{j}, EX_{i,t-1}^{j}\right) \tag{2.8}$$

$$IM_{i,t}^{j} = f\left(PR_{i,t}^{j}, DU_{i,t}^{j}, IM_{i,t-1}^{j}\right) \tag{2.9}$$

where $ST_{i,t}^{j}$, $EX_{i,t}^{j}$ and $IM_{i,t}^{j}$ are ending stocks, exports and imports, respectively, of crop j belonging to the culture group i in year t, $PR_{i,t}^{j}$ and $DU_{i,t}^{j}$ are the production and the total domestic use of crop j in year t.

For permanent crops (fruits, wine and olive oil), the functional forms implemented on the supply side are similar to grains and oilseeds except for those that determine land allocation. The equations for harvested area of fruits, wine vines and olive trees are written as

$$ah_{h,t} = \left(ah_{h,t-1} + nah_{h,t}\right), \quad h = 1,...,n \tag{2.10}$$

$$nah_{h,t} = f\left(p_{h,t-1}, \sum_{h=1}^{n} ah_{h,t}, V \right), \quad h = 1,...,n \tag{2.11}$$

Where $nah_{h,t}$ is the area of net new planting of the hth crop which is a function of the real price of that crop, the total land allocated to all permanent crops modelled and a vector V of exogenous variables that might have an impact on the area allocated to the crop being modelled. This set of exogenous variables would include direct policy support payments. In the case of wine vine area the vector V would include the vine abandonment premium as well as the new vine planting premium.

All livestock models share the same general modelling structure, though some minor differences may occur for individual livestock and meat sub-models. The key equation in all livestock models is that which determines the ending numbers of breeding animals, which have the following general form:

$$cct_{i,t} = f\left(cct_{i,t-1}, p_{i,t}, V\right), \quad i = 1,...,n \tag{2.12}$$

where $cct_{i,t}$ is the ending number in year t of the breeding animal type i, $p_{i,t-1}$ is the real price in year $t-1$ of the animal output from breeding animal i, and V is a vector of exogenous variables which could have an impact on the ending inventory of breeding animals modelled (such variables include the direct payment linked to the animals concerned or specific national policy instruments).

The numbers of animals produced by the breeding inventory are obtained as:

$$spr_{i,t} = f\left(cct_{i,t-1}, ypa_{i,t}\right), i = 1,...,n \tag{2.13}$$

where $spr_{i,t}$ is the number of animals produced from breeding herd $cct_{i,t}$ in year t and $ypa_{i,t}$ is the respective exogenous yield of animal per breeding animal of animal type i per year.

Within each *ith* livestock group there may be up to m categories of slaughter. The number of animals in the *ith* group that are slaughtered in the *jth* slaughter category is modelled as:

$$ktt_{i,t}^{j} = f\left(cct_{i,t}^{j}, p_{i,t}, z_{i,t}^{j}, V\right), \quad i = 1,...,n; j = 1,...,m \tag{2.14}$$

where $ktt_{i,t}^{j}$ is the number of animals slaughtered in category j of group i in year t, $z_{i,t}^{j}$ is an endogenous variable that represents the share of different categories of animals slaughtered in the total number of animals slaughtered for the respective group i, and V is a vector of exogenous variables that influence the slaughter category shares (e.g. a calf slaughter premium).

The average slaughter weight for animal type i is written as:

$$slw_{i,t} = f\left(slw_{i,t-1}, z_{i,t}^{j}, p_{i,t}, V\right), \quad i = 1,...,n; j = 1,...,m \tag{2.15}$$

while the total number of slaughtered animals is defined as:

$$ktt_{i,t} = \sum_j ktt^j_{i,t}, \quad i = 1,...,n; \; j = 1,...,m \tag{2.16}$$

Therefore, the total production of the *ith* meat, $PR_{i,t}$, is then derived as the product of average slaughter weight times total slaughter in that group $PR_{i,t} = ktt_{i,t} \cdot slw_{i,t}$.

Ending stocks of animals (breeding and non-breeding) of type i, $CCT_{i,t-1}$, is derived using the following identity

$$CCT_{i,t} \equiv CCT_{i,t-1} + spr_{i,t} + IM_{i,t} - ktt_{i,t} - EX_{i,t} - DL_{i,t}.$$

where $EX^j_{i,t}$ and $IM^j_{i,t}$ are exports and imports of live animals, and $DL_{i,t}$ are the death losses of animal type *i* in year *t*.

Total domestic demand for meat *i* is derived as the product of per capita demand for the meat concerned times an exogenous population variable. Per capita consumption of meat *i* can be written as

$$upc_{i,t} = f\left(upc_{i,t-1}, p_{i,t}, p_{k,t}, gdpc_t, V\right), \quad k,i = 1,...,n; \; k \neq i \tag{2.17}$$

where $upc_{i,t}$ is the per capita consumption of meat i in year t, $p_{i,t}$ is the real price of the *ith* meat, $p_{k,t}$ is a *n-1* vector of real prices of competing meats, $gdpc_t$ is the exogenously determined per capita real income and V is a vector of other exogenous variables that have an impact on per capita meat consumption.

The functional form used to estimate the ending stocks of meats has the same general form as that used in the estimation of the animal breeding inventories, Eq. 2.12. Similarly the specifications of the trade equations for live animals and meats follow the same general functional form used in the grains and oilseeds models, Eqs. 2.8–2.9.

Finally, among the AGMEMOD sub-models, the dairy model is arguably the most complicated. A particular feature of the dairy model is its emphasis on the allocation of milk fat and milk protein to the production of the various dairy commodities modelled. For each dairy commodity modelled, supply and utilisation is projected, as are wholesale prices at the country level and at the aggregate EU level.

The AGMEMOD dairy sub-model is comprised of several components. The first component determines milk production, milk imports and exports. The second component allocates available milk to feed use and fluid milk consumption, with total milk factory use (manufacturing milk) for further processing into dairy products determined as a balancing item. Milk yield per cow can be written as

$$ypc_t = f\left(p_t, qua_t, V\right) \tag{2.18}$$

where ypc_t is the yield of milk per cow in year t, p_t is the real price of milk, qua_t is the exogenous milk quota pertaining in the country concerned, and V is a vector of other exogenous variables that could have an impact on per cow yields of milk.

Dairy cow ending numbers can be written as

$$dct_t = PR_t / ypc_t \tag{2.19}$$

where dct_t is the ending numbers of dairy cows, PR_t is cow's milk production and other variables are as defined above.

Milk production in a given country, before milk quota abolition, is modelled as a function of the quota level and the real milk price

$$PR_t^{quo} = f\left(qua_t, p_t, V\right) \tag{2.20}$$

where qua_t is the exogenous milk quota in year t, p is the real price of milk in year t, and V is a vector of exogenous variables (such as policy variables, see Sect. 2.4) that affects milk production.

This equation implies that producers adjust their milk production to changes of the milk quota. The real price variable p_t reflects the contention that changes in the profitability of milk production influence producer decisions to under-fill or overfill quota.

Following quota abolition, with supply controls instruments removed, the main factor driving the level of milk production is the profitability of milk production, which can be represented in a stylised fashion by the price-cost ratio and the quota rents. As a result, the milk production equation under quota abolition has the following specification:

$$PR_t^{non\_quo} = f\left(p_t, V\right) \tag{2.21}$$

where $PR_t^{non\_quo}$ is the milk production under the non-quota regime in year t. To parameterise Eq. 2.21 estimates and projections of quota rents based on Réquillart et al. (2008) have been used. Further details of the AGMEMOD dairy sub-model's structure can be found in Chantreuil et al. (2008).

Prior to the ultimate abolition of milk quota in 2015, the period during which quota are gradually expanded (2009–2014) should see the profitability of milk production reduced due to lower milk prices and this may be associated with quota rents falling to zero. In the case of zero quota rents, the milk production Eq. 2.20 will be no longer valid. Instead, farmers' milk production behaviour is explained by the supply function expressed in Eq. 2.21 that will hold following the ultimate abolition of quota. However, in those member states where quota rents are non-zero during the quota expansion, quota is still binding so that the milk supply function of the form expressed in Eq. 2.20 is valid. Thus, depending on the time period and/or the country concerned, one or other of the milk supply functions are applied for the quota expansion period (2009–2014) with the lower of the two production levels (implied by Eqs. 2.20 and 2.21) determining the milk production.

Finally, the milk production equation, encompassing all possible conditions is represented by the following form:

$$PR_t = PR_t^{quo} \cdot dum\_quo + \min\left(PR_t^{quo},\ PR_t^{non\_quo}\right) \cdot dum\_po\_quo$$
$$+ PR_t^{non\_quo} \cdot \left(1 - dum\_quo - dum\_po\_quo\right) \qquad (2.22)$$

where dum_quo is a dummy for the milk quota period (1984–2008) and dum_po_quo a dummy for the quota expansion/phase out period (2009–2014).

Based on the milk production, the availability of milk for processing is derived. The amount of whole milk exports, feed use on farms (ufe_t), drinking milk use on farms (uff_t), and losses udl_t are subtracted from milk production, while whole milk imports are added to generate the factory use (ufa_t) of whole milk. The feed use of milk on farms is written as

$$ufe_t = f\left(ufe_{t-1},\ p_t,\ V\right) \qquad (2.23)$$

while fluid use is derived as the product of the population and the per capita fluid milk consumption. The per capita fluid milk consumption equation has the same form as that specified for the per capita meat consumption, see Eq. 2.17. The factory use of milk (ufa_t) is derived to balance the total milk supply and use.

As noted earlier, the AGMEMOD model allocates the fat and protein components of raw milk. First, the amount of fat and protein in the raw milk produced used in the manufacturing sector is calculated. This calculation involves a number of assumptions concerning the fat and protein content of the raw milk and the fat and protein content of the dairy commodities produced with manufacturing milk.

Once the available supplies of milk protein and fat are determined, the following step allocates the protein and fat components of total milk available to different dairy commodities. Milk protein allocated to the production of dairy commodity i is written as

$$ppc_{i,t} = f\left(ppc_{i,t-1},\ p_{i,t},\ p_{k,t},\ V\right),\quad k,i = 1,\ldots,n,\ k \neq \qquad (2.24)$$

where $ppc_{i,t}$ is the allocation of protein to the production of the ith dairy commodity in question in year t, $p_{i,t}$ is the price of dairy commodity i, and V represents exogenous variables that affect the protein allocation to commodity i. The total protein available is allocated to the production of n dairy commodities. Milk protein allocation equations are estimated for $n-1$ commodities, with the milk protein allocated to the production of the nth commodity derived as a balancing residual allocation.

The production of dairy commodities using milk protein is derived as the total milk protein allocation to that commodity divided by an exogenous technical protein content conversion factor. Given these production levels the allocation of milk fat to these products is also derived using exogenous fixed technical factors.

The allocation of milk fat to butter or other dairy products commodity is written as

$$fpc_{i,t} = f\left(fpc_{i,t-1}, p_{i,t}, p_{k,t}, V\right), \quad k,i = 1,\ldots,n, \; k \neq i \tag{2.25}$$

where $fpc_{i,t}$ is the fat allocation to the ith dairy commodity, $p_{i,t}$ is the price of dairy commodity i, and V is a vector that contains exogenous variables that affect the fat allocation to commodity i. Given the allocation of milk fat to butter and to other dairy products commodity, the allocation of the remaining milk fat is derived as a residual using the milk fat supply and use identity.

2.2.3 Price Formation and Market Closure

The formation of prices and market closure complete the description of the general structure of any multi-commodity country model within AGMEMOD as depicted in Fig. 2.1.

There are two kinds of price formation equations in the AGMEMOD commodity models. The first kind, which is the most common, is a *price transmission equation*, where the price in a given country is driven by an external price. In many instances this external price is the EU key price. In some other cases, for instance in the AGMEMOD oilseed models, there is no EU key price and therefore the price transmission equation relates prices in the given country with the world price and other world market indicators.

The second kind of price formation equation occurs when, for the commodity under consideration, the given country is the most important market for that commodity in the EU (e.g., France for soft wheat). In this case, the country model includes a *key price formation equation*. This equation determines the price to which price transmission equations in other country models are linked and it also "captures" all exogenous variables affecting price formation and the dynamic structure of the AGMEMOD model at the EU combined level. In particular the world market price, price policies (e.g. intervention prices) and trade agreements are included in the key price formation equation, thus indirectly affecting all country prices via price transmission equations. In addition, the key price formation equation may include as a determinant the EU self-sufficiency rate, thus making the key price (and other linked prices) responsive to the EU supply and use balance of the commodity concerned.

Therefore, when the national market is not designated as the key EU market, the price transmission equations used in the model is written as:

$$p_{j,t} = f\left(Kp_{j,t}, p_{j,t-1}, ssr_{j,t-s}, Kssr_{j,t-s}, V\right) \tag{2.26}$$

where $p_{j,t}$ is the national price of commodity j in year t, $Kp_{j,t}$ is the key price of that commodity in year t, V a vector of exogenous variables which could have an impact on the national price, $ssr_{j,t-s}$ is the self sufficiency ratio (domestic use divided by production) for commodity j in the country concerned, and $Kssr_{j,t-s}$ is the self-sufficiency rate for the same commodity in the key price market. The self-sufficiency rate in the price equations may be either current ($s = 0$) or lagged ($s = 1,2...$). The choice is made during the estimation stage and is left to the responsible provider of the country model. For its own country model each partner decides whether to include the current or the lagged self-sufficiency according to the respective quality of the estimation results.

When the national price is the AGMEMOD model key price, however, the price transmission equation is replaced by the price formation equation that is written as:

$$Kp_{j,t} = f\left(Wp_{j,t}, EIp_{j,t}, Kp_{j,t-1}, Essr_{j,t-s}, V\right) \tag{2.27}$$

where Wp_t^j is the corresponding world price, EIp_t^j is the corresponding EU intervention price, $Essr_{j,t-s}$ is the EU self-sufficiency rate for commodity j, and V is a vector of variables (exchange rates, for instance) which affects the key price and includes relevant trade policies (such as tariff and tariff rate quota levels and subsidised export limits).

As evident from how the price formation is modelled (that is, they are not market closing prices), the AGMEMOD composite model requires equations that impose the market equilibrium or closure (supply equals demand) at any commodity, country and aggregate EU level. This condition implies that *production* plus *beginning stocks* plus *imports* must always balance *domestic use* plus *ending stocks* plus *exports*. Within closed economies, this supply and use balance condition is achieved by looking for the (endogenous) market price at which equilibrium occurs. AGMEMOD commodity models, however, do not represent closed economies, as the Rest of the World markets play a role on the domestic markets modelled. Therefore, for any individual country level model, when solving the EU composite model, all commodity markets close in all years by imposing the following supply and use identity:

$$PR_{i,t} + IM_{i,t} + ST_{i,t-1} \equiv DU_{i,t} + EX_{i,t} + ST_{i,t} \quad \forall\, t;\ \ \forall i = 1,\ldots,n \tag{2.28}$$

where $PR_{i,t}$ represents the production of the *ith* commodity, $IM_{i,t}$ represents the total imports, $ST_{i,t}$ the ending stocks, $DU_{i,t}$ the total domestic use and $EX_{i,t}$ the total exports. Within any commodity model, the identity (2.28) requires the selection of a closing variable whose year-by-year value is endogenously determined and generates the necessary supply and use balance given the price level. In general, the *imports* or *exports* at the country level represent these closing variables.

Within the AGMEMOD country level models, however, no distinction is made between intra and extra EU imports and exports. To ensure a closure at the composite EU model level, it is necessary to select the commodity net exports as the model closing variable. The following identity imposes the market closure in all $c = 1,\ldots,m$ countries for any *ith* commodity:

$$\sum_{c=1}^{m}\left[PR_{c,i,t} + IM_{c,i,t} + ST_{c,i,t-1} - DU_{c,i,t} - EX_{c,i,t} - ST_{c,i,t}\right] \equiv 0 \qquad (2.29)$$

which is simply the summation over all n EU country commodity supply and use identities. Equation 2.29 can be rewritten as:

$$PR_{EU,i,t} + ST_{EU,i,t-1} - DU_{EU,i,t} - ST_{EU,i,t} - NETEX_{EU,i,t} \equiv 0 \qquad (2.30)$$

where $NETEX_{EU,i,t}$ is the net export of the EU with the Rest of the World. Thus, for the *ith* commodity model closure at the EU level is ensured by imposing the following identity:

$$NETEX_{EU,i,t} \equiv PR_{EU,i,t} + ST_{EU,i,t-1} - DU_{EU,i,t} - ST_{EU,i,t}$$

2.3 Estimating and Testing

Within the AGEMOD modelling approach, all of the behavioural, parametric relations (i.e. excluding identities), as generically presented in Sects. 2.2.2 and 2.2.3, are estimated from historical time series data. This makes the AGMEMOD model very different when compared to conceptually similar models whose parameters are often calibrated. The OECD-FAO model is one of the more prominent of such, so called, synthetic models (OECD 2008; Van Tongeren et al. 2001).

Considering the large number of countries and commodities included in the AGMEMOD model, the amount of parameters and equations to be estimated is very large. A decentralised estimation strategy is employed so as to make the task manageable and better informed: all equations of a country model are estimated by the AGMEMOD partner responsible for that country. It is thus impossible to report in detail all the estimation procedures that have been followed as well as all estimation results. Rather we summarise the general criteria provided to each AGMEMOD national team undertaking the processes of estimating the parameters of their country model. Country level econometric estimation results, detailed specification test information and complete documentation of the AGMEMOD country models can be requested, country-by-country, directly from the respective national teams. A complete list of the national teams and contacts can be found at the AGMEMOD website, http://www.agmemod.eu/.

Adherence to these general rules ensures a minimum quality standard of econometric results across country and commodity models. The general criteria break the model parameterisation process into a maximum of five stages: pre-estimation, estimation, post-estimation, calibration and validation.

2.3.1 Pre-estimation

The *pre-estimation* step in passing from the depicted country model structure to the model estimation concerns the selection of the parametric equations to be estimated and their functional form specification. *Selection* simply means the exclusion of identities from the estimation stage, while *functional specification* means the transformation of the implicit form $y = f(x, z)$ into an explicit parametric structure. For econometric reasons and, above all, to avoid computational complexity at the stage of model combination and closure, all the parametric equations in the AGMEMOD model are linear in the unknown parameters. They can also be linear in the explanatory variables, but this is not strictly necessary. In fact, variables can enter the equations as ratios, square roots, quadratic and cubic terms, or logarithmic transformations.

Finding the appropriate functional specification, however, implies the understanding of which additional regressors (besides those just mentioned) may improve the estimation quality and thus would better capture the real underlying relations among the variables. These regressors mostly concern time dummies, time trends, lagged dependent or independent variables as well as variable transformations (squared, logarithmic variables, etc.). Such variables may be particularly helpful as time series data are in use. In such case, the data generation processes should be investigated to assess the appropriate functional specification to be adopted. This investigation process includes the testing for autocorrelation, structural breaks and non-linearities. Nonetheless, additional variables can be very helpful in limiting the econometric consequences of these time series properties. Adding a *time trend* can be helpful for apparently or nearly unit-root processes. *Time dummies* can take into account non-stationarity or non-linearity that is generated by the presence of one or more structural breaks. *Transformed* (as *nth*-order polynomials, for instance) or *interaction terms* can account for non-linearities or more complex relations, while *lagged variables* can be useful in addressing problems with autocorrelation. For more details and clarifications on these econometric issues, the reader is directed to Stock and Watson (2010) and Greene (2007) (for introductory and advanced econometrics, respectively) and Enders (2009) (for time series econometrics) textbooks.

The general objective of the pre-estimation stage is to achieve a linear regression model that is correctly estimable using *ordinary least squares* (OLS) methods. In some instances, the time series properties of the regressor and/or regressand may make the OLS estimator inconsistent. Nonetheless, the inclusion of additional variables such as outlined above may be useful in restoring the consistency of OLS estimation and, thus, in achieving reliable estimation results.

2.3.2 Estimation

Once an appropriate linear regression specification has been chosen for all model equations, perhaps through the introduction of the appropriate regressors, they can be estimated by appending a conventional identically and independently distributed error terms ~ $N(0,\sigma^2)$. This *estimation stage* is mostly based on least squares estimation. The choice has been made to avoid other potential estimators (*maximum likelihood, ML,* or *generalised methods of moments, GMM,* estimators; e.g. Greene 2007; Stock and Watson 2010). Some of these alternative estimators would actually ensure consistency and higher efficiency than OLS estimators. Nonetheless, the estimation stage would become more complex and time-consuming. Moreover, these estimators could imply heterogeneity in estimation strategies across different country models (for instance, for the selection of instruments in the *GMM* case). Finally, the results from non-OLS estimators might not necessarily be qualitatively better than OLS results due to unclear small-sample performances.

For these reasons, OLS estimation is the default estimation procedure over all AGMEMOD estimated equations. This choice does not mean, however, that other Least Squares (LS) estimators are disregarded whenever they are explicitly called for by the model's structure. In particular cross-equation relations require system rather than equation-by-equation estimation in order to obtain consistent and efficient estimates. Therefore, the first step of the estimation stage consists in identifying equations that can be estimated singularly with OLS and those equations that require LS system estimators (Greene 2007). Three different types of system estimations can be applied to the AGMEMOD structure.

First, some equations are recursive: i.e. left hand side variables in one equation enter as right hand side variables in other equations, but *vice versa* does not hold. A possible alternative to separate OLS estimation is to estimate the endogenous variables at first, and then substitute these with their estimated values in the second equation, akin to an *Instrumental Variables-two stage least squares (IV-2SLS)* estimation strategy. In practice, however, the depicted structure of the model recursiveness often implies the presence of lagged endogenous variables among regressors. In such circumstances, the single-equation OLS estimation remains appropriate.

Secondly, there are equations with simultaneity: left hand side variables in one equation enter as right hand side variables of other equations and *vice versa*. Here, single-equation OLS estimation leads to possible endogeneity bias. The alternative solution is a *three stage least squares (3SLS)* system estimation approach, that may restore consistency and efficiency. Such cases, however, are quite rare in the AGMEMOD modelling structure. They occur in some country crop models where the harvested area equation depends on yields and the yield equation itself depends on the cultivated area.

Thirdly, there are equations that are apparently unrelated (no endogeneity or recursiveness occurs), but in which cross-correlation of the respective error terms occurs. This typically occurs in demand system estimation. In such cases, instead of applying a single-equation OLS estimate, the *Seemingly Unrelated Regressions* (SUR) estimation approach can significantly improve the quality of parameter estimates (Greene 2007).

The adoption of the more suitable system estimation instead of single-equation OLS estimation thus depends on the particular model specification at the country level. Therefore, any AGMEMOD partner responsible for a country model is also responsible for the choice of the appropriate estimation procedure.

The estimation stage ultimately provides point estimates and associated standard errors for all model parameters. These latter usually require some further evaluation and elaborations.

2.3.3 Post-estimation

The *post-estimation* stage concerns the evaluation of the estimated equation/models' goodness of fit and the consequent possible feed-back impact on the overall model performance. In this respect, the minimum requirement in AGMEMOD before proceeding from the estimation to validation stage involves the achievement of the expected signs for parameters associated with key model variables, such as prices and policy measures. A second requirement is, beside the sign, that these estimated parameters be statistically different from zero. Consequently, t-tests on all estimated parameters are systematically run. Conventional specification tests (*F, LM, LR* or *Wald*) (Greene 2007) are performed when an equation does not respect both requirements on sign and significance of key parameters. These tests assess whether the introduction or exclusion of regressors (such as dummies, trends, quadratic or interaction terms) improve the statistical quality of the estimate. Despite attempts to improve the equations statistical performance some parameter may remain statistically insignificant, but with signs as expected.

2.3.4 Calibration

The implication of statistically insignificant parameters for the overall model performance, however, has to be carefully evaluated and this evaluation may lead to the use of calibration in the determination of model parameters. The *calibration* of specific model parameters is most often necessary where short and incomplete data series prevent the use of regular estimation procedures. In such cases, some parameters are chosen on the basis of estimates available from the literature, expert knowledge or results obtained from similar equations in other country models. The remaining parameters, such as the constant term and possibly time trend parameters are estimated so as to fit the equation as much as possible to the statistical data that are available.

2.3.5 Validation

To this point the outlined estimation and testing procedures concerned single model equations or limited systems of equations. The testing and *validation* of the entire model, however, is a key step in achieving a simulation model. This stage is especially important in evaluating the model's internal consistency, dynamic properties and in-sample predictive quality. To validate AGMEMOD, within-sample simulations, ex-post policy evaluations and shock analyses are applied to individual country models and to the combined model.

Within-sample simulation results often indicate the necessity to revise the estimation procedure for particular model equations. They provide important information about the qualitative performance of estimated equations within the overall model context. When all model equations have been (re-)estimated and the model (including identities and model closure equations) has been set up in the chosen programming environment, within-sample simulation experiments are conducted. Historically observed values are used as values of the exogenous model variables. As the observed values of all endogenous variables in the sample are known, these can be compared with the model's predicted values. This makes it straightforward to calculate within-sample fits for individual model equations and for the whole model. Measures such as the *mean absolute percentage error* which is a measure of predictive quality, and the *mean percentage error* that provides an overall measure of the model's projection error are used to evaluate model performance. Based on these diagnostic measures, the dynamic within-sample performance of the model equations is evaluated and equations that give rise to problems are identified and selected for re-estimation.

Ex-post policy evaluations are applied to examine the correctness of the model's theoretical structure, empirical model parameterisation and dynamic adjustment mechanisms in response to changes in exogenous variable values. Such analysis is not only helpful in evaluating and comparing outcomes with intended effects, but it also provides guidance to improving the model's performance. Ex-post policy analyses are mostly conducted by assessing counterfactual policy scenario experiments aimed at answering questions such as: *What would have happened if a certain policy had not been implemented?* Such experiments are applied under the ceteris paribus condition, which assumes that everything else – except the evaluated policy instruments – remains the same.

Shock impacts are used to test the entire model's performance. How does the model respond to shocks to exogenous variables such as exchange rates, population growth rates or world market prices? To evaluate the dynamic properties of the model, selected exogenous variables are shocked one-by-one and the simulation experiments are run over a long time period. The magnitudes and direction of the simulated effects are evaluated against a priori or theoretical expectations. This will lead to a re-estimation process where inconsistency with the economic theory occurs.

2.4 Policy Modelling

In this section we show how the heterogeneity of the CAP is accounted for as part of the AGMEMOD PE modelling framework. Given that the AGMEMOD model's parameters are largely based on econometric estimates the AGMEMOD's evaluation of policy changes is based on the reaction of agri-food markets to policy and market changes during the sample period over which the model's parameters were estimated. When the original AGMEMOD model was developed (Riordan 2005), the main analytic focus was the response of agricultural supply and demand to changes in key European market prices and to changes in the value of coupled direct payments. Given recent and ongoing CAP reform processes and the increasingly heterogeneous implementation of the CAP in EU member States (see Chap. 1) a richer approach to the incorporation of CAP policy instruments was developed as part of the FP6 project (Salputra et al. 2011).

Beginning with the MacSharry reforms of 1992, the CAP evolved with a focus on production related direct support (payments per area and per animal head). Up until 2004, the modelling approach used to examine CAP support under Agenda 2000 was in general also appropriate for the evaluation of policies in the new Member States that had acceded to the EU from 2004. In these countries pre-accession support was mostly coupled to agricultural production, crop area or animals. Following the 2003 Fischler reform and the enlargement of the EU in 2004, direct income support to farmers became available without an obligation to produce a specific volume of production. A methodological approach to modelling such policy instruments, termed the *policy harmonisation* approach, has been developed as part of the AGMEMOD project to capture the effects generated by different policy measures (Salputra et al. 2011). This approach allows for the systematic examination of the impact of existing direct support policies as well as those proposed and that might be expected in the future.

While market price support and supply control in the EU are based on the implementation of a homogenous systems across Member States, the direct income support system of the CAP is particularly diverse in both design and implementation. In line with the CAP Health Check decisions (Council Regulations (EC) No. 72/2009, 73/2009 and 74/2009), the diverse agricultural policy systems permissible under the CAP may be gradually harmonised over the period 2010–2013. Firstly, through the mandatory decoupling of the major part of direct payments that under the Fischler reform could be retained as coupled by Member States and secondly, through the voluntary switch from historically established payments to regional flat payments in EU15 Member States.

Currently, there are many direct payment schemes, which can be broadly categorised into five groups:

- *historical scheme*: individual farmers get the amount of direct income support they obtained in a reference period. The payments are unequally distributed among farmers on the basis of their activities in the reference period and the differing levels of direct payment support across farming activities. Some of the payment schemes can also remain directly coupled, which directly affects the production decisions;
- *regional flat area payment scheme*: regional amounts (budgetary envelopes) are divided among the number of eligible hectares declared in the region. Some of the Agenda 2000 payment schemes can remain coupled;

- *static hybrid scheme*: a combination of historical rights and regional flat area payments. These models lead to the redistribution of budgetary support between the sectors though this redistribution is not as strong as in the case of the regional flat area payment scheme. Some of the payment schemes can also remain coupled as well;
- *dynamic hybrid scheme*: the share of the budgetary resources shifts from a historical to a regional flat area basis over the period to 2013, while the share of resources devoted to historical payments tends towards zero;
- *simplified area payment scheme*: this flat area payment system is applied in most new Member States. From an economic perspective, this system is similar to the regional flat area payment scheme, with the principal difference that coupled payments are restricted to specific support measures, as set out in Article 68(1) of Regulation (EC) No. 73/2009. EU budgetary resources that are devoted to the simplified area payment scheme are gradually phased-in over the period to 2013 up to the level that new Member States are allowed to top-up from national funds.

Table 2.1 presents the direct payment schemes as implemented in the different EU Member States in 2010. Partially coupled payments are still applied in one third of the Member States, while specific support and other kind of coupled support schemes (for producers of rice, starch potatoes, protein crops, nuts, seeds, cotton, sugar, fruits and vegetables) are implemented in almost all countries.

Figure 2.2 illustrates the allocation of the Pillar I funding to different direct support schemes (coupled, historical and regional). Due to the phasing-in of direct payments in the new Member States, the total funding as well as the share of regional payments in their payment envelopes will increase over the period to 2013.

To make the AGMEMOD model capable of analysing the consequences of switches in agricultural policy regimes, all applicable direct support measures have been implemented within AGMEMOD model's policy block. Within this policy structure, all budgetary support granted to the agricultural sector is accounted for on a systematic basis according to the relevant Council regulations. The links between the different policy measures in the model ensure the consistent consistent of policy effects in case of a switch between policy schemes and changes in policy objects.

The implementation of the policy harmonisation approach in the AGMEMOD model involves the following three steps:

- the collection of country specific policy information about all types of direct payments, which is detailed enough for analysing CAP impacts on the supply of agricultural products;
- the construction of consistent country datasets showing the coherence between different types of EU CAP direct support elements in their allocation and providing the input for policy modelling (see Chap. 3 for greater detail);
- the elaboration of the block of policy variables that is included in the general modelling structure.

The complexity of allocating direct supports funded from the EU budget and distributing them according to common rules requires standardised and centralised

Table 2.1 Direct payment schemes in Member States, 2010

Direct payment scheme	Country	Partially coupled payments for cattle, sheep, goats	Specific support	Other coupled support
Historical scheme	Belgium	+	+	+
	Ireland		+	+
	Greece		+	+
	Spain	+	+	+
	France	+	+	+
	Italy		+	+
	Netherlands		+	+
	Austria	+	+	+
	Portugal	+	+	+
	UK (Scotland, Wales)			+
Regional flat area payment	Malta			
	Slovenia	+	+	+
Static hybrid scheme	Luxembourg			
	Sweden	+	+	+
	UK (Northern Ireland)			+
Dynamic hybrid scheme	Denmark	+	+	+
	Germany		+	+
	Finland	+	+	+
	UK (England)			+
Simplified area payment scheme	Bulgaria		+	+
	Czech Republic		+	+
	Estonia		+	
	Cyprus			+
	Latvia		+	+
	Lithuania			+
	Hungary		+	+
	Poland		+	+
	Romania		+	+
	Slovakia		+	+

calculations within the combined AGMEMOD model structure. The policy harmonisation approach, by systematising and harmonising the use of agricultural policy data across the country models, ensures that the implementation of policy across different country models is comparable and thus facilitates the consistent modelling of policy at and EU levels.

The methodology used to classify the direct support measures under different CAP payment systems follows the *Producer Support Estimate* component definitions of the OECD (2008):

- *coupled support*: defined as payments based on output, on area planted or animal numbers; production is required;

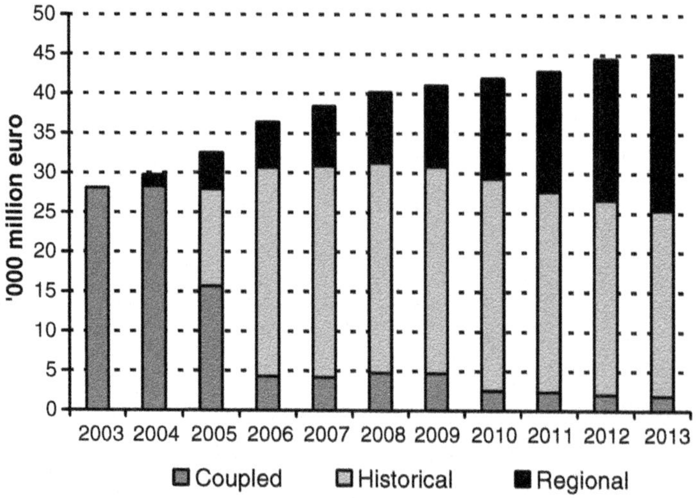

Fig. 2.2 Coupled, historical and regional direct payments (Source: calculations based on EC Agriculture and Rural Development DG "Agriculture in the European Union – Statistical and economic information" 2003–2009; Regulation (EC) No 1782/2003; Regulation (EC) No 552/2007; Regulation (EC) No 73/2009)

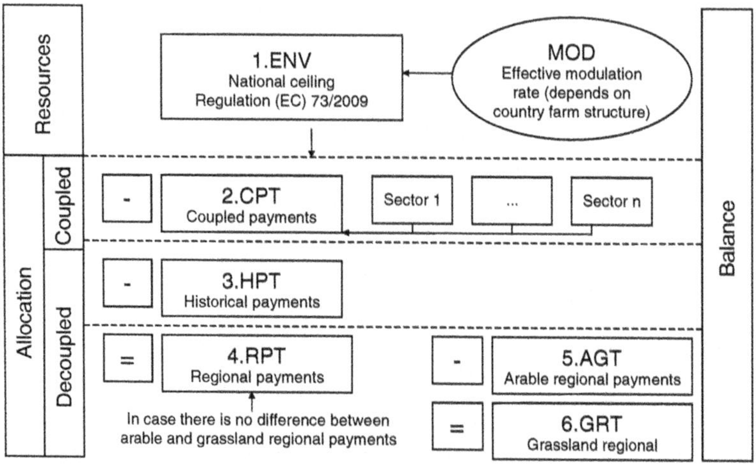

Fig. 2.3 Allocation of national budgetary ceilings

– *decoupled support*: defined payments Member State based on non-current area and animal numbers; production is either required or not required (historical entitlements).

Figure 2.3 illustrates the allocation of Member State's national direct payment envelopes (budgets) over the various different agricultural direct payments types.

The effective national envelope of the Pillar I direct payments env_t equals the national ceiling after adjustment for the deduction that arises from the application of the effective modulation rates.

$$env_t = ENV_t \cdot (1 - cmo_t - vmo_t) \tag{2.31}$$

where ENV_t is the national ceiling defined by regulations, cmo_t is the compulsory modulation rate and vmo_t is the voluntary modulation rate. The compulsory and voluntary modulation rates have been calculated on the basis of existing farm subsidy structures in each of the Member States.

The ceiling for the total coupled payments envelope, cpt_t, is determined by com_j^p, the premiums p for commodity j, by the maximum coupling rates defined for each premium and commodity, crt_{tj}^p, by the country specific reference number for premium p and livestock commodity j, ref_j^p, and by the reference yield and reference area for crop commodity j, ydr_j and ahr_j^p:

$$cpt_t = \sum_{p=1}^{s} \left[\sum_{j=1}^{l'} crt_{tj}^p .com_j^p .ref_j^p + \sum_{j=l'+1}^{l} crt_{tj}^p .com_j^p .ydr_j .ahr_j^p \right]. \\ \left(1 - cmo_t - vmo_j\right), \text{where } j = 1,\ldots,l',l'+1,\ldots,l;\ p = 1,\ldots,s \tag{2.32}$$

where the total number of commodities, l, is distinguished between l' livestock products and $(l\text{-}l')$ crops. Decoupled historical payments hpt_t are calculated using the following formula:

$$hpt_t = ENV_t \cdot hrt_t \cdot (1 - cmo_t - vmo_t) \tag{2.33}$$

where hrt_t is the share of historical payments in the national ceiling chosen by each country. In addition, the following formula is used for the decoupled regional payments rpt_t:

$$rpt_t = env_t - cpt_t - hpt_t, \forall t \tag{2.34}$$

The policy harmonisation approach allows for the quantitative assessment of the impact of various elements of the EU CAP direct support schemes. All direct payments are recalculated and treated as a policy price add-on to the relevant producer price to form a *reaction price*. The reaction price increases the margin between production returns and costs. In this way CAP direct income supports affects EU supply and demand balance, the external trade and ultimately market prices.

The reaction price accounts for the effect of decoupled direct payments through the use of *multipliers*, which adjust the share of budgetary support that is reflected in that commodity's reaction price. It is assumed that the support coupled to a product or a production factor associated with a particular product directly impacts production. Also the support granted to land, irrespective of the type of product produced, can stimulate production. The magnitude of the multipliers applied to different types of decoupled subsidies depends on the nature of the support payments. Based on OECD studies (OECD 2006), the value of the regional multiplier is set to 0.3, while the value of the historical multiplier is set to 0.5. The historical payments lead to a greater production incentive than the regional payments because the appropriate production technologies have already been established on the farms. In the case where payments are fully coupled to production, the multipliers are set to 1. Decreased production incentives have been found to be associated with decoupled direct payments by Gohin (2006), Rude (2008) and Balkhausen et al. (2008).

The reaction prices, when deflated by the appropriate input cost indices, are the economic variables that drive the supply decisions of the farmers within the AGMEMOD model's structure. Changes in decoupled payment values lead to responses by farmers that are analogous to, but smaller than, farmers' responses to changes in agricultural output prices.

Within the AGMEMOD model, reaction prices for commodity *j* are simulated as endogenous variables and adjusted on the basis of assumptions concerning the value of policy input variables (modulation rates, coupling rates, multipliers and variables controlling the allocation of budgetary envelopes between coupled payments, and regional and historical payment schemes) and endogenous commodity market prices. Equations 2.35 and 2.36 express the relationship between the policy price add-ons prc_{tj} and the exogenous policy variables in the AGMEMOD crops, meat and milk sub-models respectively.

$$prc_{t,j} = \left(\frac{cpm \cdot cpt_{t,j}}{\max\left[ahr_{t,j}, ah_{t-1,j} \right]} \right) + \left(\frac{hpm \cdot hpt_t + rpm \cdot rpt_t}{ah_{t-1}} \right) \cdot \left(r_{t-1,j} \right)^{-1} \tag{2.35}$$

$$prc_{t,j} = \left(\frac{cpm \cdot cpt_{t,j}}{cct_{t-1,j}} \right) + \frac{\left(\frac{hpm \cdot hpt_t + rpm \cdot rpt_t}{ah_{t-1}} \right)}{utr_{t-1}} \cdot \left(slw_{t,j} \right)^{-1} \tag{2.36}$$

Here *cpm*, *hpm* and *rpm* are the multipliers for coupled, historical and regional payments respectively that determine the impact of different direct payments types on the reaction prices to which farmers respond when making their production decisions. The variable $r_{t-1,j}$ in Eq. 2.35 reflects the yield of crop *j*, while the variable utr_{t-1} in Eq. 2.36 is the lagged average livestock density per hectare.

The sum of policy price add-ups (prc_j) and producer prices (p_j) reflect the reaction prices, which are then used as explanatory variables in the supply side equations of the AGMEMOD country models (see Sect. 2.2.2).

2.5 Conclusion

This chapter illustrates the first conceptual steps underlying the AGMEMOD approach. It consists in building the AGMEMOD model as an aggregation across countries and commodity markets of single-country and single-commodity models. Moreover, it describes the general structure of these single-country and single-commodity models. It then provides a template of the AGMEMOD model's behavioural equations and summarises the guidelines followed in AGMEMOD to define appropriate functional specifications and, eventually, to estimate these equations.

The actual modelling and estimation work o at the single country level has, using the general structure and the common guidelines, been carried out by the respective AGMEMOD partners. Due to obvious space limits, this chapter does not detail the model specification and estimation country-by-country and commodity-by-commodity.

This model construction work is the basis of the next two steps of the AGMEMOD conceptual modelling framework. In the second step the country models are aggregated to form the EU AGMEMOD composite model. To achieve this, once estimated, the country models are transferred to an appropriate suitable computer format supplemented with a functional user interface. This transition from the conceptual model, as described in the present chapter, to the computer model is illustrated in the next Chap. 3.

Third, once the conceptual model has been implemented in a working computer model, projections can be generated. This is achieved by building and running scenarios. Within AGMEMOD, building scenarios involves the definition of alternative projections of the exogenous variables that underlie the model and, then, forcing the model to generate projections of the endogenous variables, conditional on the revised policy or other exogenous data. There are three principal sets of exogenous data within AGMEMOD: macroeconomic variables, external prices (mostly world prices) and policy variables. The construction and the generation of scenarios are presented in Chaps. 4 and 5. In the latter case projections of alternative policy scenarios are based on a series of different assumptions concerning EU agricultural policy in the period 2014–2020.

The AGMEMOD approach to modelling policy instruments, termed in this chapter as the policy harmonization approach, allows for the quantitative assessment of the impact of various elements of EU CAP direct support schemes. The production effects of direct payments (coupled and decoupled) are accounted for through reaction prices formed by the addition of producer prices and policy price add-ups. By systematising and harmonising the management and use of policy data, the policy harmonisation approach allows for the examination of the impact of existing direct support policies as well as those proposed and expected in the future.

References

Balkhausen O, Banse M, Grethe H (2008) Modelling CAP decoupling in the EU: a comparison of selected simulation models and results. J Agr Econ 59:57–71

Chantreuil F, Hanrahan KF, Levert F (2005) The Luxembourg agreement reform of the CAP: an analysis using the AG-MEMOD composite model. In: Arfini F (ed) Proceedings of the 89th EAAE seminar "Modelling agricultural policies: state of the art and new challenges". Monte Università Parma, Parma

Chantreuil F, Donnellan T, Van Leeuwen M, Salamon P, Tabeau A, Bartova L (2008) EU Dairy quota reform – AGMEMOD scenario analysis. Paper presented at the 12th Congress of European Association of Agricultural Economists, Ghent, Belgium, 26–29 Aug 2008

Enders W (2009) Applied econometric time series, 3rd edn. Wiley, New York

Gohin A (2006) Assessing CAP reform: sensitivity of modelling decoupled policies. J Agr Econ 57(3):415–440

Greene WH (2007) Econometric analysis, 6th edn. Prentice Hall, Upper Saddle River

Hanrahan KF (2001) EU GOLD model manual. Mimeo Rural Economy Research Centre, Teagasc, Dublin

OECD (2006) Decoupling agricultural support from production. Policy Brief, OECD, Paris. Available via http://www.oecd.org/dataoecd/5/54/37726496.pdf. Cited 21 June 2011

OECD (2008) OECD-FAO Agricultural Outlook 2008–2017. OECD and FAO. OECD Publishing, Paris

Réquillart V, Bouamra-Mechemache Z, Jongeneel R, Penel C (2008) Economic analysis of the effects of the expiry of the EU milk quota system. Report prepared for the European Commission, IDEI, Toulouse

Riordan B (2005) Building local knowledge into EU agri food projections: experiences of a fifth framework co-ordinator. EuroChoices 4(1):32–37

Rude J (2008) Production effects of the European Union's single farm payment. Can J Agr Econ 56(4):457–471

Salputra G, Chantreuil F, Hanrahan KF, Donnellan T, van Leeuwen M, Erjavec E (2011) Policy harmonized approach for the EU agricultural sector modelling. Agr Food Sci 20(2):119–130

Stock JH, Watson MW (2010) Introduction to econometrics, 3rd edn. Addison-Wesley, Reading

Van Tongeren FW, Meijl H, Surry Y (2001) Global models applied to agricultural and trade policies: a review and assessment. Agr Econ 26(2):149–172

Chapter 3
AGMEMOD Model

Myrna van Leeuwen, Foppe Bouma, Frédéric Chantreuil, Wietse Dol, Emil Erjavec, Kevin F. Hanrahan, Petra Salamon, and Guna Salputra

Abstract This chapter presents the AGMEMOD computer framework, introduces the reader to the available software and explains how to use this software. The technical aspects of building the AGMEMOD database, the generation and use of the computer model and the solving of the AGMEMOD model are presented in detail. This chapter can be used as a guideline for readers who are interested in the operational version of the AGMEMOD model.

Keywords Database • Modelling framework • GAMS • User interface • Validation

The purpose of this chapter is to describe how AGMEMOD's conceptual model structure, as presented in Chap. 2, has been transferred into a computable framework. This computer framework must allow the following objectives to be met:

- to integrate all Member State models into a combined EU version with particular attention to the role of price linkages and the model closures;

M. van Leeuwen (✉) • F. Bouma • W. Dol
LEI Wageningen UR, Alexanderveld 5, The Hague 2585 DB, The Netherlands
e-mail: myrna.vanleeuwen@wur.nl

F. Chantreuil
Université de Caen Basse Normandie, UFR Sciences Economiques et Gestion, 17 rue Claude Bloch, BP 5186, F-14032 Caen cedex, France

E. Erjavec
Biotechnical Faculty, University of Ljubljana, Groblje 3, 1230 Domžale, Slovenia

K.F. Hanrahan
Rural Economy Research Centre, Teagasc, Athenry, Co Galway, Ireland

P. Salamon
Johann Heinrich von Thünen-Institute (vTI), Brunswick, Germany

G. Salputra
Latvia State Institute of Agrarian Economics, 14 Struktoru St, Riga, Latvia

F. Chantreuil et al. (eds.), *The Future of EU Agricultural Markets by AGMEMOD*,
DOI 10.1007/978-94-007-2291-0_3, © Springer Science+Business Media B.V. 2012

- to provide a transparent, flexible and consistent modelling tool;
- to make baseline projections up to a 10 years period at the country and EU27 levels;
- to carry out simulations and to analyse results.

We have built the AGMEMOD computer framework using the General Algebraic Modelling System (GAMS) language (McCarl 2010). The AGMEMOD computer framework has been designed and tailored to the needs of the AGMEMOD model user and hence provides a certain level of flexibility to the user.

This chapter introduces the reader to the AGMEMOD software and explains how to use this software (Sect. 3.3). The main technical aspects of building the AGMEMOD database (Sect. 3.1), and building and using the AGMEMOD computer model are also covered in detail (Sect. 3.2). The generation of the model database and computer model is closely related to the earlier steps undertaken in specifying the functional modelling structure of the AGMEMOD model and in the estimation and calibration of the complete equation set. The compilation and successful simulation of the computer model depends on the successful combination of earlier work described in Chap. 2 and the procedures described in this chapter.

This chapter can be used as a guideline for readers who are interested in using the operational version of the AGMEMOD model. Files, programs and tools mentioned are part of the AGMEMOD 4.0 version. A demo version of the AGMEMOD model is available at Springer's Extra Materials website http://extras.springer.com/.

3.1 Database

The development of the modelling database is a key task in building the AGMEMOD modelling framework. Figure 3.1 represents the general procedure that is applied in the database building process for countries and the EU, it also shows the interaction of database tasks and other processes related to AGMEMOD model development (Chantreuil and Levert 2007).

The data demands of the AGMEMOD modelling approach are high, as time series for parameter estimation purposes are required not only for the supply side of agricultural commodity markets but also for the different type of uses and processing demands. Each country model is based on a database template of annual time series, which depending on the country concerned ranges from as early as 1973 to 2006. AGMEMOD's database is in part made up of supply and use balance sheets for all commodities. These include data on opening stocks, production, imports, human food consumption, feed use, processing and industrial use, exports, and ending stocks, at the level of primary agricultural commodities and, often, also their first processing level. These commodity balance variables, together with commodity prices, are determined inside the model and belong to the set of *endogenous variables* of the AGMEMOD model. Where possible the AGMEMOD Partnership uses Eurostat sources such as the Agricultural Information System (Verhoog 2000) and NewCronos to populate this database.

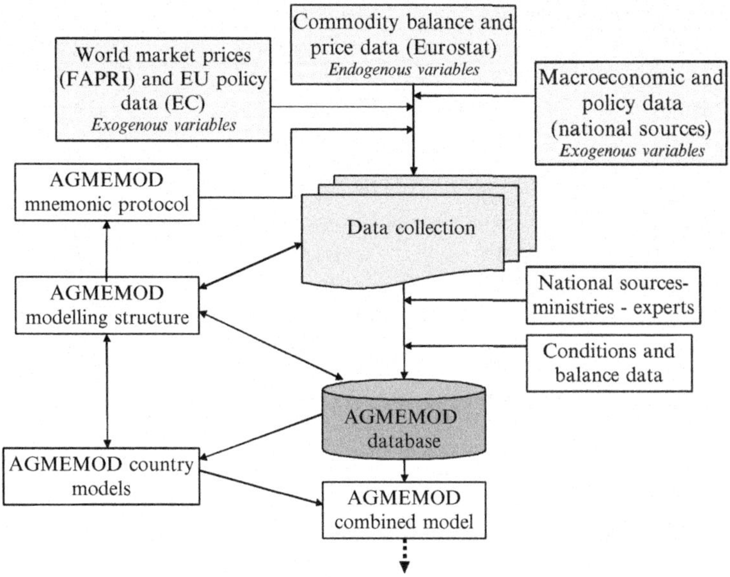

Fig. 3.1 The AGMEMOD database

A second subset of data captures the evolution of CAP policy instruments, such as direct payment instruments and income supports. In the projection period these are either held constant at the politically agreed levels (baseline) or altered in line with the analysed policy change scenario. A third subset of data covers macroeconomic variables such as population, inflation, per capita GDP, and currency exchange rates and their exogenous projections. These data are obtained from the national statistical services in the Member States or from internationally recognized macroeconomic forecasters. A fourth subset of data captures world market price projections, which are in the main obtained from the FAPRI modelling system (FAPRI 2010). World market prices, policy instruments and macroeconomic data are determined outside the model and belong to the *exogenous variables* of AGMEMOD.

3.1.1 Mnemonics Protocol

A mnemonic protocol must ensure that each country model and its constituent data and parameters satisfy the technical requirements for the combination within a multi-country modelling framework (Hanrahan 2001). The protocol outlines the rules and conventions regarding the *mnemonics* or *naming* of commodities, activities and countries used in the AGMEMOD model. Variable names consist of seven or eight letters and are presented as 2-3-2 or 2-4-2 combinations. The first part indicates the *commodity (group) code,* of which Table 3.1 presents a selection.

Table 3.1 Selection of commodity codes used in AGMEMOD

Commodity	Mnemonic	Commodity	Mnemonic
Soft wheat	WS	Cattle and calves	CC
Barley	BA	Hogs and pigs	HP
Maize	CO	Beef and veal	BV
Rapeseeds	RS	Pig meat	PK
Sunflower seed	UF	Sheep	SH
Soybeans	SB	Broilers	BR
Potatoes	PT	Ewes	EW
Sugar beets	ST	Poultry	PO
Sugar	SU	Cow's milk	CM
Apples	AP	Butter	BU
Tobacco	TB	Cheese	CD
Cotton	CT	Skim milk powder	NF

Table 3.2 Selection of activity codes used in AGMEMOD

Activity	Mnemonic	Activity	Mnemonic
Production	SPR	Area harvested	AHA
Domestic consumption	UDC	Yield per hectare	YHA
Imports	SMT	Slaughtering weight	SLW
Exports	UXT	Nominal price	PFN
Ending stocks	CCT	Wholesale price	PWM
Beginning stocks	ITT	Cost indices	ICI

Table 3.3 Selection of country codes used in AGMEMOD

Country	Mnemonic	Commodity	Mnemonic
Austria	AT	Latvia	LV
Belgium	BE	Lithuania	LT
Bulgaria	BG	Netherlands	NL
Czech Republic	CZ	Poland	PL
Denmark	DK	Portugal	PT
Estonia	EE	Romania	RO
France	FR	Spain	ES
Finland	FI	Slovenia	SI
Germany	DE	United Kingdom	UK
Greece	GR	Turkey	TR
Hungary	HU	Macedonia	MK
Ireland	IE	Croatia	HR
Italy	IT	Russia	RU

Similarly, Table 3.2 provides a selection of *activity codes*, while Table 3.3 shows a list of *country codes* indicating, respectively, the second and third parts of the permitted variable names. A complete list of mnemonics used in the AGMEMOD model is available in Hanrahan et al. (2006) and at Springer's Extra Materials website http://extras.springer.com/.

Through the application of the AGMEMOD mnemonic protocol all endogenous and exogenous variables used in the AGMEMOD model have a unique mnemonic code. For example, the variable name WSSPRFR denotes *soft wheat* (WS) *production* (SPR) in *France* (FR), while BVPFNHU depicts *beef and veal* (BV) *nominal price* (PFN) in *Hungary* (HU).

3.1.2 Endogenous Variables

The maintenance and updating of the AGMEMOD datasets, in an internally consistent and coherent manner, are important prerequisites for the ongoing use of the AGMEMOD model as a policy analysis tool. As noted earlier, each AGMEMOD country model is based on an aligned database that contains annual time series for agricultural commodity supply and uses balance sheets and price data related to the commodities modelled.

The initial stage of the data collection process often encounters problems associated with incomplete or inconsistent data, and unbalanced supply and use data. Given the political upheavals of the last 30 years, for some Member States the range of time series data can be very short and the definition of data reported can be variable.

The AGMEMOD model, as an equilibrium model, requires that supply and demand databases are balanced for all commodity markets for all years, and for all countries modelled. Where supply and use data are not balanced, partners have to seek additional country information from national statistical institutes, research institutes, industrial organizations and expert knowledge, with the aim of improving the data consistency and quality and ensure that the necessary balances hold. Several basic rules are followed in order to ensure balanced and consistent databases for each year, commodity and country. These rules are recommendations, which are as follows:

- keep data for production and domestic consumption as close as possible to their values in the original dataset;
- values of all supply and use elements, with the exception of changes in stocks, must be non-negative;
- keep recent variables close to their observed values;
- recalculate missing or negative values as an average of the preceding and following years data points;
- adjust *stocks, imports* and *exports* to satisfy the supply and use balance or add a *losses/statistical discrepancy* variable so as to ensure supply and use balance.

This second stage of the data collection process is central to the construction of complete, convenient and consistent country and EU databases. The satisfaction of balance conditions within the AGMEMOD country and EU databases can be explained by using as an example of how the land area harvested and crop markets are related in the AGMEMOD model. Figure 3.2 illustrates the process of allocatin of *Total land area* (TLAHA) over the different land area types collected by Eurostat.

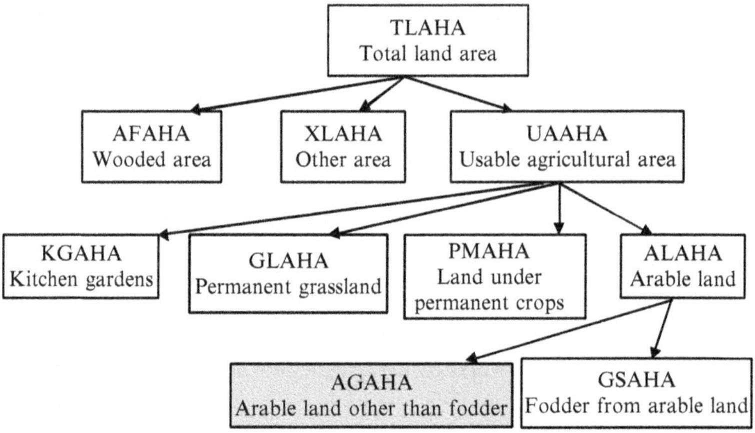

Fig. 3.2 Land area type relations

Table 3.4 Conditions on land area types

Condition 1	TLAHA = UAAHA + AFAHA + XLAHA
Condition 2	UAAHA = ALAHA + GLAHA + PMAHA + KGAHA
Condition 3	**AGAHA** = ALAHA − GSAHA

The figure indicates that the data series for *Arable land other than fodder* (AGAHA, grey box in Fig. 3.2) is derived from the *Arable land* (ALAHA) and *Fodder from arable land* (GSAHA) variables in the Eurostat database. Table 3.4 notes the identities that must hold in each country database. In all of following tables and figures the colour grey has been used to identify the variables that are derived from an equality condition.

Figure 3.3 summarizes the way in which *Arable land other than fodder* (AGAHA) has been built up from different categories of crop land uses. Table 3.5 denotes the identities that must hold in the country's databases.

Figure 3.4 shows how soft wheat supply and demand are related to *total grains area* (GRAHA) Table 3.6 lists the soft wheat market identities that must hold in all country databases, for all years.

For each commodity market modelled, equivalent flow diagrams and supply and use balance conditions have been determined in order to provide a complete and comprehensive market database. Given the large number of crops and livestock markets in AGMEMOD, the number of conditions that must hold in each country database is very large. See Chantreuil and Levert (2007) for a complete list of all balance conditions.

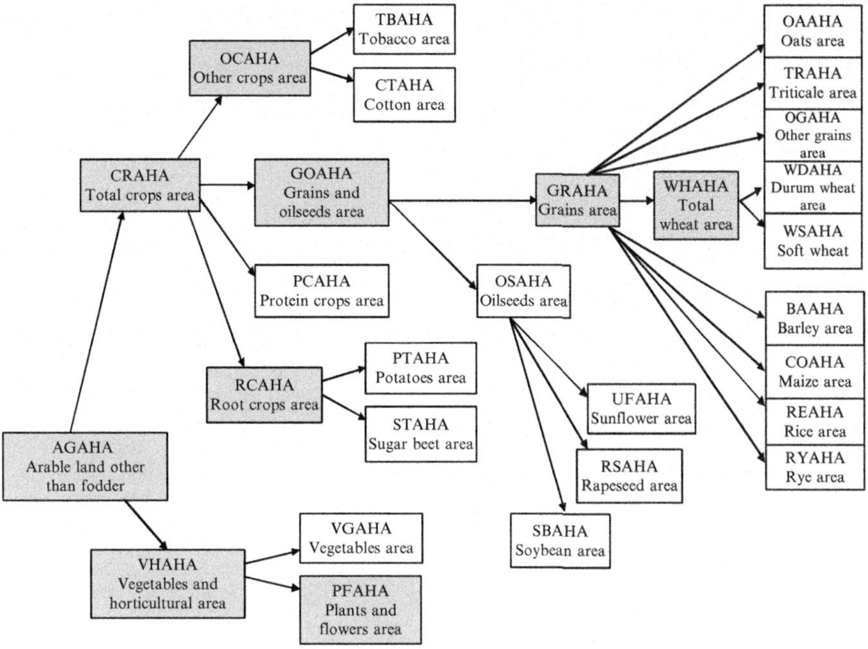

Fig. 3.3 Arable land area categories

Table 3.5 Conditions on arable land area categories

Condition 4	**AGAHA** = **CRAHA** + **VHAHA**
Condition 5	**CRAHA** = **RCAHA** + PCAHA + **GOAHA** + **OCAHA**
Condition 6	**OCAHA** = TBAHA + CTAHA
Condition 7	**GOAHA** = GRAHA + OSAHA
Condition 8	**GRAHA** = OGAHA + WHAHA + BAAHA + COAHA + REAHA + RYAHA + TRAHA + OAAHA
Condition 9	WHAHA = WSAHA + WDAHA
Condition 10	**OSAHA** = UFAHA + SBAHA + RSAHA
Condition 11	**RCAHA** = STAHA + PTAHA
Condition 12	CFAHA = ORAHA + **XCAHA**
Condition 13	**VHAHA** = VGAHA + **PFAHA**

3.1.3 Exogenous Variables

Data on the exogenous variables used in the AGMEMOD model are by definition determined outside of the AGMEMOD model and reflect information on world market prices, agricultural and trade policy, and country level macroeconomic data.

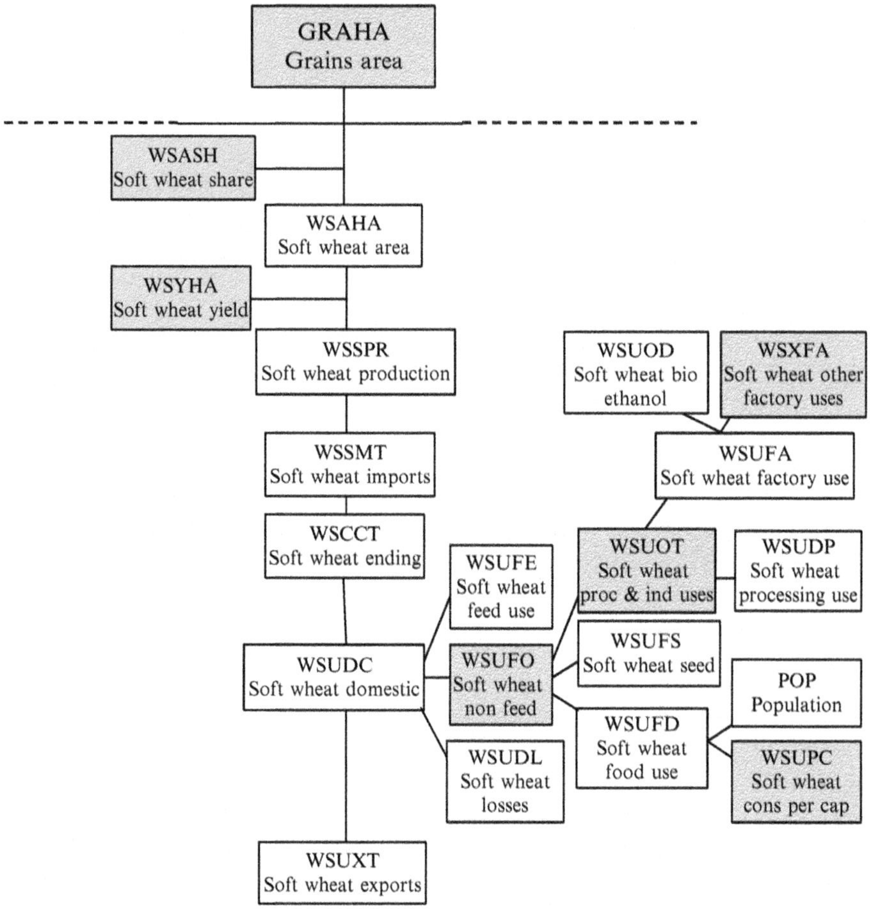

Fig. 3.4 Soft wheat flow diagram

Table 3.6 Conditions on soft wheat market

Condition 14	WSAHA = GRAHA * WSASH
Condition 15	WSSPR = WSAHA * WSYHA
Condition 16	WSUDC = WSUFO + WSUFE + WSUDL
Condition 17	WSUFO = WSUOT + WSUFS + WSUFD
Condition 18	WSUOT = WSUFA + WSUDP
Condition 19	WSUFA = WSUOD + WSXFA
Condition 20	WSUFD = WSPOP * WSUPC
Condition 21	WSSPR + WSCCT(−1) + WSSMT = WSUDC + WSUXT + WSCCT

Historical data on these variables are necessary to empirically estimate the models of supply and utilization at the country level. In order to simulate the AGMEMOD country and EU levels and to generate baseline and scenario projections to a 10 year horizon, medium term projections of all exogenous variables are also required.

Table 3.7 World market prices (USD/tonne) used in AGMEMOD

Mnemonic	Description	Source
WHPMDUS	Wheat price, U.S. Gulf	FAPRI
BAPMDUS	Barley price, U.S. Portland	FAPRI
COPMDUS	Maize price, U.S. Gulf	FAPRI
RSPMDUS	Rapeseed price	Oil World, FAO
SBPMDUS	Soybean price	Oil World, FAO
UFPMDUS	Sunflower seed price	Oil World, FAO
SUPMDUS	Sugar (raw), NY 96%, fob Caribbean ports	FAPRI
CCPNDUS	Steers price, Nebraska	FAPRI
HPPLDUS	Hogs, U.S. 51–52% lean	FAPRI
BRPWDUS	Broiler price, U.S. 12-city	FAPRI
SHPWDNZ	Lamb price, New Zealand	OECD
NFPMDNL	Skim milk price, FOB North Europe	FAPRI
CDPMDNL	Cheese price, FOB North Europe	FAPRI
BUPMDNL	Butter price, FOB North Europe	FAPRI
CLPMDUS	Cotton lint price	FAPRI
OLPMDUS	Olive oil price	FAO
TOPMDUS	Tomato price	FAO

3.1.3.1 World Market Prices

In general, AGMEMOD incorporates historical and 10 year world market price projections series from the FAPRI world agricultural modelling system (FAPRI 2010). World prices for oilseeds, lamb and tomatoes are taken from Oil World and the annual OECD-FAO agricultural market outlook (OECD-FAO 2010). World livestock and grain prices are market prices from the US, whereas dairy commodity prices and oilseed, oilseed meal and oil prices are generally northern European prices. Table 3.7 provides a selection of the mnemonics and sources of the world market prices used in AGMEMOD model.

3.1.3.2 Macroeconomic Data

AGMEMOD also requires data on a range of macroeconomic variables to empirically estimate the multi-commodity and multi-country models. Historical data on macroeconomic variables like population, inflation, per capita income levels and currency exchange rates are assembled at the country level. Macroeconomic projections, to a 10 year horizon, are obtained from European Commission services and from national statistical and economic research services. These are used together with the estimated model parameters and the model databases to generate baseline and scenario outlooks. Projections are checked by partners to ensure that radically divergent projections for the development of inflation, currency exchange rates and economic growth across EU Member States are avoided.

Table 3.8 Direct policy instruments in old Member States

Period	Source of policy instruments
……–1998	*EUROSTAT-Economic Accounts for Agriculture*: product subsidies paid
……–2006	*OECD*: national direct support
1999–2006	*EC, DG-Agri*: budgetary expenditures for direct payments, number of application for premiums, CAP support measures
2007–2020	*EC Regulations* (EC R1782/2003, EC R73/2009): budgetary national ceilings, reference level of production eligible for support and premiums
	EC DG-Agri: CAP support, country decisions

Table 3.9 Direct policy instruments in new Member States

Period	Source of policy instruments
……–2003	*EUROSTAT-Economic Accounts for Agriculture*: product subsidies paid
2004–2006	*OECD* and *national data sources*: direct payments, national top-ups
1999–2006	*OECD*: national direct support, budgetary expenditures
	EC DG-Agri: CAP phasing-in support
2007–2020	*EC Regulations* (EC R1782/2003, EC R73/2009): budgetary national ceilings, reference level of production eligible for support and premiums
	EC DG-Agri: CAP support, country decisions

3.1.3.3 Market Support Policy Instruments

To analyse the impact of agricultural policy on agricultural markets within the context of an econometrically estimated partial equilibrium model framework requires detailed historical policy data. The AGMEMOD policy dataset captures the evolution of the CAP over the period 1973–2006 using data obtained from the European Commission services and the OECD. The database incorporates data on the level and value of direct payment instruments and support prices associated with the commodity market organizations that collectively make up the CAP. Trade policy data on commodity export subsidies and tariff rate quota have been collected from the OECD. This EU policy dataset is used in the development and estimation of the AGMEMOD country models described in Chap. 2.

3.1.3.4 Direct Support Policy Instruments

While the market support policy and supply controls in the EU are based on the implementation of a homogenous system across EU Member State countries, the direct income support scheme as introduced from 2005 onwards under the Fischler CAP reform is particularly diverse (see Sect. 2.4).

The transition from a policy based on market price support to decoupled direct income support schemes via coupled direct payments systems (see Chap. 2) requires the collection of additional policy data such as national envelopes, reference areas and livestock numbers, direct payment rate per reference unit, maximum coupling rates for products and Member State modulation rates. Tables 3.8 and 3.9 list the required policy data and their sources in the historical period for old and new EU Member States. For the projection period such policy variables are set equal to their politically agreed level, or varied according to the definition of the policy change scenario being analysed.

The *national budgetary ceilings* for direct payments are based on the EU funding for agricultural direct support as established by EC R1782/2003 (European Council 2003) and EC R73/2009 (European Council 2009). There are three sets of *budgetary envelopes for direct coupled payments* that, depending on the Member States concerned and their implementation of the CAP, may be applicable. These are:

- the budgetary ceilings for direct payments partially coupled, which are relevant for countries that have implemented the SPS;
- the budgetary ceilings for coupled direct payments under the Classical scheme, which are relevant for old Member States and Slovenia;
- the budgetary ceilings for coupled direct payments in the pre-accession period or other envelopes not included in national ceilings (top-ups in SAPS and other direct aid schemes), which are relevant for new Member States other than Slovenia.

There are two groups of budgetary envelopes for direct payments to be granted as decoupled payments:

- *budgetary historical payments*: single payments implemented under the historical, static hybrid or dynamic hybrid schemes that move to a flat rate SP scheme, or regarding the sugar, fruit and vegetable payments in new Member States;
- *budgetary regional flat rate payments* derived envelopes for SAP or SP schemes.

The conditions that must hold with respect to these budgetary envelopes were set out in Sect. 3.5.

Modulation is the CAP policy instrument that redistributes funding from the CAP Pillar I to CAP Pillar II. In the old Member States, the incidence of modulation at the farm level depends on each farm's total direct income support received from Pillar I. On average, the effective modulation rates in each Member State will vary as a function of the country's farm structure and in the AGMEMOD model this is used to determine the share of Pillar I payments in old Member States that are modulated. In the new Member States, the modulation mechanism will come into force after the phasing-in period of CAP direct income supports is completed in 2013. From 2014, the calculated average effective modulation rates per country are based on the farm support structures in 2008 in the Member State concerned. Figure 3.5 depicts the modulation rates that are applied in the AGMEMOD model and which redistribute national ceilings between Pillar I and Pillar II in each of the EU27 Member States.

The area of arable crops planted, and cattle and sheep farmed, for which CAP premiums were granted in old Member States, have until 2004 been based on a *reference period* and payment quotas. The policy reference level of areas, yields, number of animals, amount of production and the maximum rates of premiums for CAP direct payments are considered for the time period since 2005.

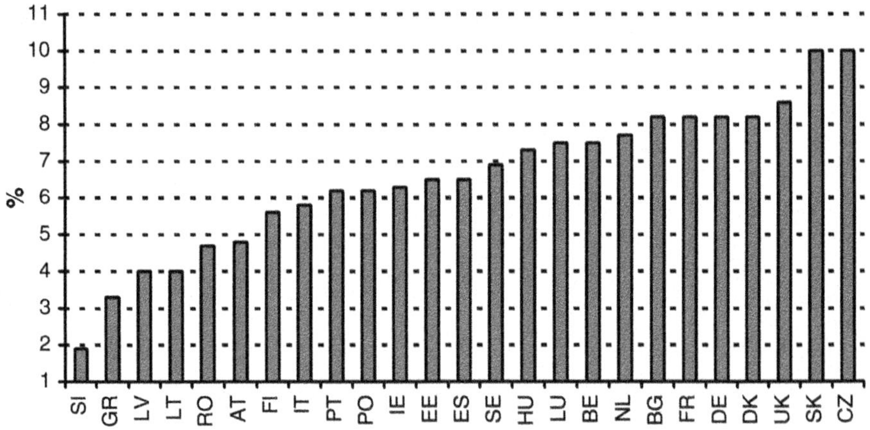

Fig. 3.5 Average effective modulation rate in Member States, 2013 (Source: Based on indicative figures on the distribution of aid and size-class of aid, received in the context of direct aid paid to the producers according to Council Regulation (EC) No 1782/2003 (Financial Year 2008))

The maximum *coupling rate* levels of direct payments are set out in Regulation (EC) No 1782/2003. Corrections regarding the application years after the introduction of the CAP Health Check have been based on Regulation (EC) No 73/2009. Decisions on the coupling rates for old Member States and Slovenia take account of those direct payments that remain coupled and the implementation of CAP reform in the tobacco, cotton, olive oil, hops and sugar sectors. The applied coupling rates for new Member States under the SPS from 2013 have been based on the implementation of the CAP as agreed in the recent CAP Health Check reform (see Chap. 2 for further detail).

3.1.4 Data Generator Tool

The large number of commodity markets that are modelled in AGMEMOD, the large number of balance conditions that must be satisfied, and the necessity to frequently update country databases, motivated the development of a *data generator* tool. This tool, based on GAMS and MS-Excel software, allows for a quasi-automatic construction of commodity balance sheets at the country level. Figure 3.6 illustrates the role of the data generator tool in the AGMEMOD endogenous country data update process.

As noted earlier, the AGMEMOD country database is mainly based on Eurostat sources. To ensure a complete and coherent database that satisfies the AGMEMOD mnemonic protocol as well as the imposed commodity balance conditions, the data

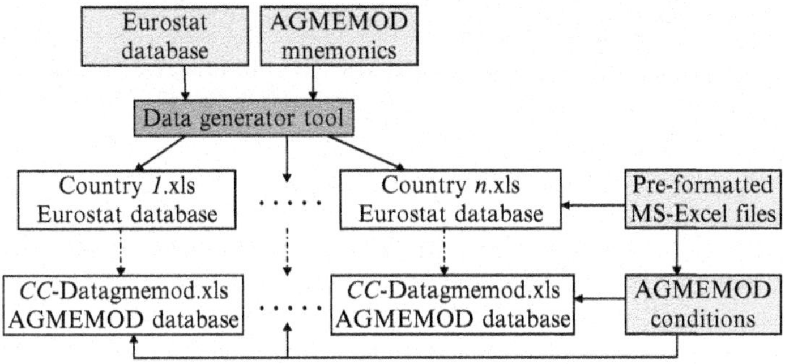

Fig. 3.6 Data generator tool for endogenous variables

generator tool transforms Eurostat datasets into compatible country data files (where the file for country CC is named *CC-datagmemod.xls*). The corresponding *CC-Datagmemod.xl*s file is comprised of four spreadsheets:

- *Eurostat*: this sheet contains the Eurostat data for country *CC*, automatically filled by the data generator tool;
- *Database*: this sheet contains the checked, adjusted and completed *CC* country database;
- *Eurostat balance checking*: this sheet automatically applies and checks the equilibrium conditions with the *Eurostat* sheet;
- *Database balance checking*: this sheet automatically applies and checks the equilibrium conditions with the *Database* sheet.

The *Database* sheet is the end result of the second stage of the data collection work done by partners in case the automatically collected Eurostat data show imbalances or missing values. The *Database* sheet contains the complete and consistent country database that is used for the AGMEMOD model.

The *Eurostat balance checking* sheet and the *database balance checking* sheet use a colour code to check if a given condition is satisfied or not. A cell is highlighted in green if the corresponding condition is respected, and it is highlighted in red if the necessarycondition does not hold. A country database is only considered as complete and ready for used within the AGMEMOD model when all cells on the *database balance checking* sheet are green coloured.

Each variable of the AGMEMOD model is allocated to a unique parameter type. Column 1 of Table 3.10 provides the available parameter types, while column 2 denotes whether they are endogenous or exogenous to the model. The third column indicates that the endogenous variables are stored in the *CC.datagmemod.xls* and the exogenous variables are located in either the files *AssumptionsInput.xls* or *PolicyHarmonisation.xls.*

Table 3.10 Parameter types and associated data files used in AGMEMOD

Parameter type	Description	Data file
V2(P_A, C,T1)	Product P and Activity A in Country C in year T1; *endogenous* to the model	CC-datagmemod.xls
VWP(P_A,T1)	World market price for Product P and Activity A in year T1; *exogenous* to the model	AssumptionsInput.xls (*world* sheet)
VPOL(P_A,T1)	EU dependent policy variable for Product P and Activity A in year T1; *exogenous* to the model	AssumptionsInput.xls (*EU-policy* sheet)
VPOLC(P_A,C,T1)	Policy variable for Product P and Activity A in Country C in year T1; *exogenous* to the model	AssumptionsInput.xls (*CC-policy* sheet)
VPH(P,A,C,T1)	Policy harmonization variable for Product P and Activity A in Country C in year T1; *exogenous* to the model	PolicyHarmon.xls
VMAC(A,C,T1)	Macroeconomic variable for Activity A in Country C in year T1; *exogenous* to the model	AssumptionsInput.xls (*macro* sheet)
TREND(T1)	Trend variable in year T1, *exogenous* to the model	AssumptionsInput.xls (*trend* sheet)
DUM(D,T1)	Dummy D in year T1; *exogenous* to the model	AssumptionsInput.xls (*dummy* sheet)

3.2 Combination of Country Models

A bottom-up approach is used to integrate country models into the AGMEMOD combined EU model. Chapter 2 described how country models have been specified and econometrically estimated on the basis of commodity model templates. Adherence to these commodity model templates in the different country level AGMEMOD models makes is possible to meaningfully combine the different country models while still allowing each country model to reflect (via their parametric structures) the heterogeneous situations of the different Member State's agricultural systems. The maintenance of this analytical consistency across the many country models is essential to facilitating the meaningful comparison of policy impacts across the different countries modelled within the AGMEMOD model.

3.2.1 From Conceptual to Computer Model

The building of models and the writing of properly structured software with which to simulate these models must go hand in hand. If simulation software is poorly structured, the process of making changes to models can become prone to error and excessively time consuming. Properly organized and documented model simulation software contributes to a model's flexibility, extendibility, reproducibility and transferability. In general, simulation models such as AGMEMOD tend to be adjusted frequently as they are used to answer new research questions; this potentially ad hoc process leads to new model versions. Overtime, due to the accumulation of

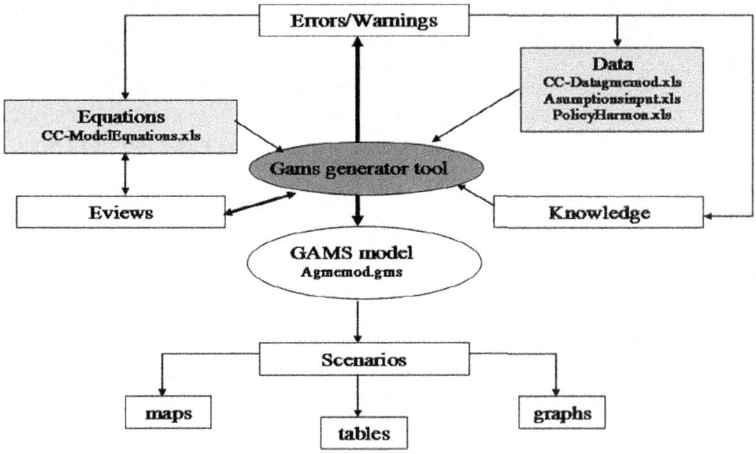

Fig. 3.7 Procedure from data handling to scenario analysis

changes to model structures and parameters the consistency between conceptual model and actual computer model can become less than clear. To reduce the incidence of such problems, both conceptual model builders and IT-scientists work together in the AGMEMOD project to develop and maintain a flexible AGMEMOD computer framework, that:

- has a clear set-up;
- is flexible enough to meet requirements of various research projects and allow for the development of different model versions;
- is easily extendable with new commodities;
- is easily extendable with new countries;
- is easily extendable through the addition of new analytical tools;
- provides reliable results, which are reproducible.

Figure 3.7 illustrates the global procedure applied to the AGMEMOD model from the data preparation, the estimation of equations and the generation of GAMS framework to model solving and scenario analysis processes. The grey boxes refer to the input needed from modellers such as the assembly of data (Sect. 3.1) and the estimation of equations (Chap. 2), while the ovals refer to the AGMEMOD software that has been developed to guide and assist these modelling processes. In general, the procedure works as follows. First, all common exogenous data (stored in *AssumptionsInput.xls* and *PolicyHarmon.xls*) and specific country data (stored in *CC-Datagmemod.xls*) are read to create a comprehensive GAMS dataset that is used to solve the combined model. Second, the dataset is integrated with the estimated country level equations (stored in *CC-ModelEquations.xls,* see Sect. 3.2.2). Then, solutions for all markets, years and countries are generated using GAMS and the endogenous model results are exported to output files. These model output files capture the endogenous projections of agricultural activity levels (areas harvested, livestock numbers), supply and use balances (production, domestic use, imports, exports and ending stocks) and prices at the country and EU levels.

The *GAMS generator* tool, which is in the centre of the figure, plays a role as the mediator between the development of the conceptual model and the development of the computer model. This use of this tool is explained in more detail in Sect. 3.2.2.

3.2.2 GAMS Generator Tool

Before the AGMEMOD computer model can be built in GAMS code, all estimated and calibrated country equations must be specified in a MS-Excel file, named *CC-ModelEquations.xls*. AGMEMOD distinguishes three types of equations:

- *EQ* type (stands for *Eq*uation), which indicates that the variable is estimated or calibrated;
- *IDEN* type (stands for *Iden*tity), which indicates that the variable is defined as a calculation of other variables;
- *FX* type (stands for *F*i*x*ed), which indicates that the variable is fixed on its last observation value and is thus exogenous to the model.

Each equation of the AGMEMOD model has to be defined as one of these three equation types.

Secondly, to achieve a consistent AGMEMOD modelling framework, the existence of a one-to-one relation between the three databases (*CC-Datagmemod.xls*, *AssumptionsInput.xls* and *PolicyHarmon.xls*) and the estimated equations (*CC-ModelEquations.xls*) across the individual country models must be ensured. The GAMS generator tool has been developed with the objective of guaranteeing the creation of consistent, transparent and error free GAMS programs. In this way, the GAMS generator tool forms the bridge between data and estimated equations used on the one side and the GAMS model to be generated on the other. It ensures the achievement of consistent and transparent GAMS country models in the sense that requirements on the use of time indices, bounds and parameter types are all fulfilled automatically. The main objectives of the GAMS generator tool are as follows:

- to check that all variable data have been implemented in the data files (*CC-Datagmemod.xls*, *AssumptionsInput.xls* and *PolicyHarmon.xls*) where these variables have been specified in the equation file (*CC-ModelEquations.xls)*;
- to check that "foreign" data (i.e. key price data from other countries) used in the own country models are available in the corresponding "foreign" *CC-Datagmemod.xls*;
- to check the use of mnemonics and the type of equations;
- to report errors and problems regarding the AGMEMOD database and equations;
- where Eviews (2011) is used to estimate country models, the GAMS generator tool can be used to re-estimate complete country models;

– to transfer the estimated parameters from Eviews into GAMS code, with inclusive sets for products, activities, countries and time.

To summarize, there are two conditions that must be fulfilled in order to ensure a consistent AGMEMOD computer framework. First, the relationship between the number of variables and data series used in the data files and the number of variables and equations used in the equation file *(CC-ModelEquations.xls)* must be linear. Second, the variables used in the specified equations of *CC-ModelEquations.xls* must exactly fit with the unique mnemonic codes and dimensions of the corresponding variables in the data files. Both conditions are crucial for ensuring that the GAMS generator tool can automatically transfer the mnemonics into the correct GAMS parameter types. To support this process, the GAMS generator tool provides an error and warning list when the basic conditions are not fulfilled and will only produce the GAMS code needed to run the AGMEMOD model when all errors have been solved.

AGMEMOD Partners are free to choose which econometric or statistical software to use in estimating or calibrating their AGMEMOD country models. In general, two ways of filling the *CC-ModelEquations.xls* file have been followed:

– based on *EViews* econometric software (EViews 2011). As this is the most commonly used econometric software in the AGMEMOD partnership, specific AGMEMOD software was developed to transfer information from EViews files to the *CC-ModelEquations.xls* files automatically;
– based on other econometric software packages, e.g. GRETL (Cottrell and Lucchetti 2007). In this case, partners individually managed the transfer of information from associated estimation files to the *CC-ModelEquations.xls* file.

3.3 AGMEMOD User Interface

AGMEMOD has been built with a flexible and transparent modelling structure (M'barek and Bartova 2007; Van Leeuwen et al. 2008). The entire system operates under a GAMS user interface, which contains options to read and revise endogenous and exogenous data, to read and revise estimated equations, to solve and re-solve the model, to generate baseline and scenario projections and to present model output. This section describes how these steps can be managed through the AGMEMOD user interface. Readers who are interested in using the demonstration version of the AGMEMOD model can use this section as guide. The AGMEMOD demonstration version is available at Springer's Extra Materials website http://extras.springer.com/,

Figure 3.8 shows the opening screen of the AGMEMOD user interface with menu options relating to reading in of data, the model solving process, output presentation, and documentation in the upper tool bar.

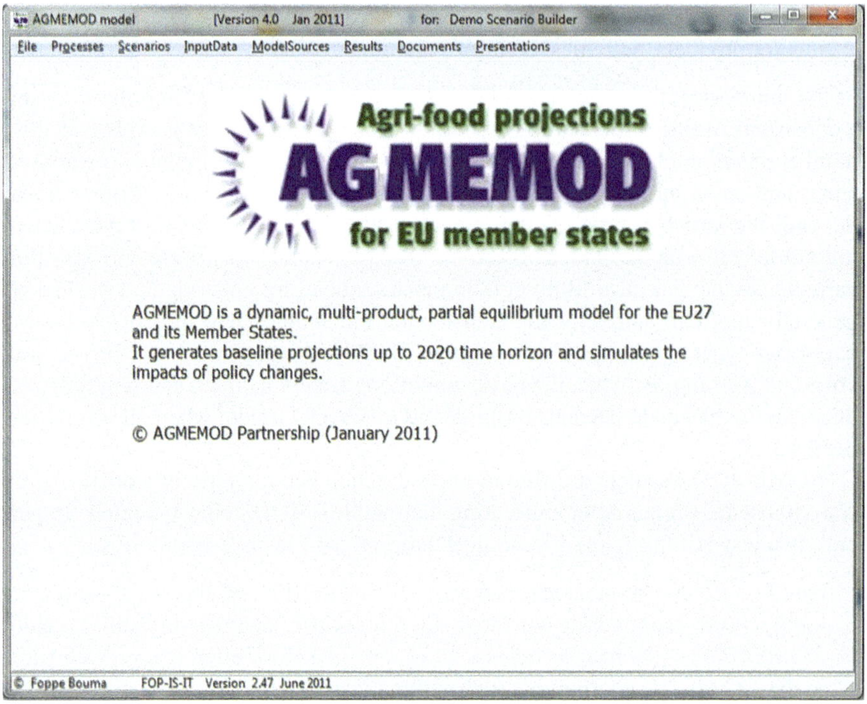

Fig. 3.8 Menu options of AGMEMOD user interface

3.3.1 *Viewing and Revising Data*

As already described, the AGMEMOD model uses three MS-Excel data files that contain endogenous and exogenous data at the country and EU levels. These files are:

- *CC-Datagmemod.xls*: observed commodity balance and price variables;
- *AssumptionsInput.xls*: observed and projected macroeconomic variables;
- *PolicyHarmon.xls*: observed and projected policy variables.

In order to reduce the number of data files that are imported into the GAMS model, the data series contained in each *CC-Datagmemod.xls file* are transferred into an associated *CC* sheet of the *HistoryData-EU27.xls* file.

Besides these three files, the AGMEMOD model needs the file *CountryTimeSet.xls*. This file contains information on the range of the historical and simulation time periods per country modelled.

Generally data files are revised as and when better or updated data become available or when additional data for new commodities or new countries are implemented within the AGMEMOD model. To explore and edit historical data that are stored in *HistoryData-EU27.xls,* use the menu option *InputData* in the AGMEMOD user interface:

InputData\AGMEMOD Excel input data\History data-countries (*HistoryData-EU27.xls*)

This file includes commonly structured worksheets (indicated by Country mnemonics) for each country that belongs to the AGMEMOD model. Historical data series can be revised or extended at the country level. Depending on the data availability in Eurostat and national data sources, the end year of the observed data series may differ between countries. Consequently, the start year of the simulation period across countries may also vary. The AGMEMOD model deals with any differences in the length of country data series in a flexible way. This process is managed using the file *CountryTimeSet.xls*. This file is viewed and edited by selecting:

InputData\AGMEMOD Excel input files\Time sets-countries (*Countries TImeSet.xls*)

The *CountryTimeSet.xls* file guarantees that the AGMEMOD model always uses and presents the latest available observed data. Moreover, it also guarantees that the first year of the simulation period immediately follows the latest observed year for each country modelled. This is explained using an example. The range of the Netherlands AGMEMOD commodity database is 1973–2007. In this example this period will be extended with data for 2008. Thus, both the length of the historical period (which becomes longer) and the length of the projection period (which becomes smaller) will change. This information is managed through the file *CountryTimeSet.xls* as follows. First, the '*2008*' column in the '*NL*' row of the *HistoricalData* sheet must be filled with a *one*, so that the '*NL*' row will contain *ones* for the years 1990–2008. Next, the *one* under the '*2008*' column in the '*NL*' row of the *RealProjectionPeriod* sheet must be removed, so that the row will contain only *ones* for the years 2009–2020. This action ensures that the AGMEMOD model will generate projections for the Netherlands over the period 2009–2020, while it takes the observed data values as given for the 1990–2008 period.

The macroeconomic and policy assumptions that are exogenous to the AGMEMOD model can be reviewed or changed by selecting

InputData\AGMEMOD Excel input data\Macroeconomic and policy assumptions (*Assumptionsinput.xls*)

InputData\AGMEMOD Excel input data\ Policy harmonization assumptions (*PolicyHarmon.xls*)

Updates of observed and projected macroeconomic and policy assumptions are implemented via changes to these MS-Excel files. Section 3.3.4 describes how the AGMEMOD data reading process is managed using the AGMEMOD user interface.

3.3.2 Viewing and Revising Equations

To explore the model equations set for a particular country, e.g. Germany, choose the menu option *InputData* from the AGMEMOD user interface:

InputData\AGMEMOD Excel equations *(CC-Modelequations.xls)\ ..Germany*

This opens the MS-Excel file *DE_ModelEquations.xls* (*DE* is the country mnemonic for Germany) and shows a series of variables with similarly structured rows as in the *DE* sheet of the *HistoryData-EU27.xls* file. Column *H* shows the specification of the equations in terms of estimated parameter values and the explanatory variables used. The model equation file, in this example *DE_ModelEquations.xls,* is updated if variables are re-specified or re-estimated or when equations for a new variable are added to the model. The old specifications, listed under column *H,* are simply replaced with their updated version. To avoid over-specification, each unique mnemonic combination used in the equations must have a data record with the same unique mnemonic combination in one of the data input files mentioned in Sect. 3.3.1.

3.3.3 Viewing and Revising the GAMS Model

The best way to explore the GAMS model for a country (or group of countries) is by choosing the menu option *ModelSources* from the AGMEMOD user interface:

ModelSources \..2 Model solving phase \ 2.1 Agmemod GAMS model

This opens the file *Agmemod.gms* which captures the complete AGMEMOD model in GAMS code. The code of a specific country, e.g. Germany, can be selected in the tree structure on the left-hand side of the screen:

ModelForTotalEU.gms – EUMainModel.gms – CountryListAllCountryModels.gms *–Equations_DE.gms*

This action reveals the GAMS code of the German AGMEMOD model on the right-hand side of the screen. Due to the size of the German model, the file has been decomposed into sub-models, for example *Grains and Oilseeds Supply and Use*, which can easily be explored through the tree structure on the left hand side of the user's screen.

AGMEMOD partners frequently update and revise their own country data and equation sets. Without the use of the checking tools described earlier, this could lead to inconsistency and potential model harmonization problems each time that individual country models are integrated into an updated aggregate EU version. As noted in Sect. 3.2.2, the GAMS generator tool plays a key role in transferring (re-) estimated variables from the *CC-ModelEquations.xls* files into consistent and transparent GAMS code. The GAMS generator tool can be used through the AGMEMOD user interface by choosing:

ModelSources \.1 Data and Model preparation phase by using Agmemod2Gams tool \ 1.1 Adjust CC_ModelEquation file and transfer equations into GAMS code

Following this choice a screen appears, in which the country model (as represented in the *CC-ModelEquations.xls)* that is to be transferred into GAMS code is chosen. By for example selecting *DE* in the **Choose country/Country to process** box and by clicking on the **Start Processing** button, the GAMS generator tool starts to check that the following data and equation conditions:

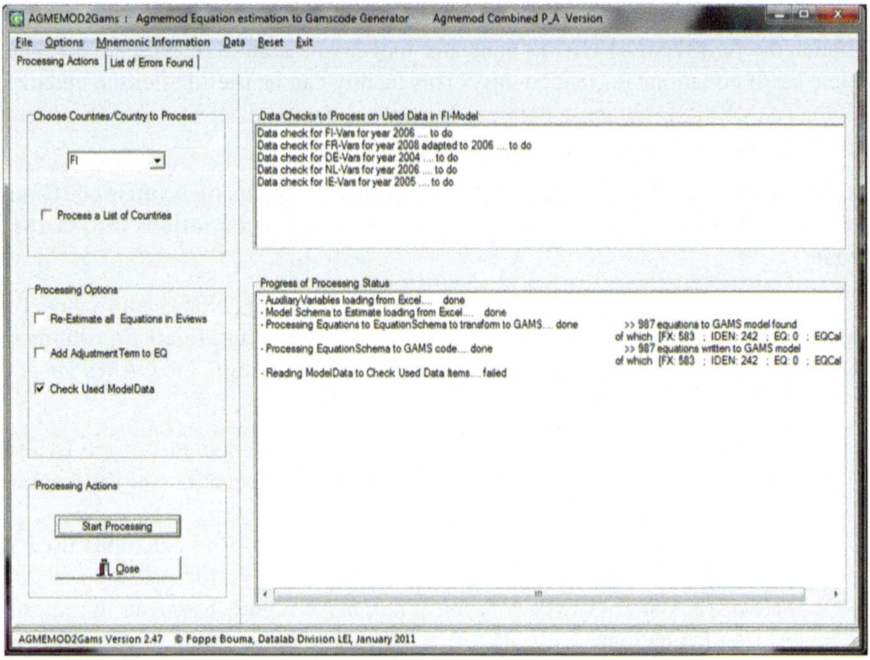

Fig. 3.9 Process to check data and equations and to generate GAMS code

- that all endogenous variables used on the right hand side of the specified model are specified in the file *DE-ModelEquations.xls*. Variables that do not fulfil this condition are listed in an error text file;
- that all endogenous variables in *DE-ModelEquations.xls* have data series in the *DE* sheet of the file *HistoryData_EU27.xls*. Variables that do not fulfil this condition are listed in the error file;
- that all exogenous variables used on the right hand side of the specified equations in the file *DE-modelequations.xls* have data series in *AssumptionsInput.xls* and *PolicyHarmon.xls*. Variables that do not fulfil this condition are listed in the error file;
- that all key prices or other "foreign" variables used on the right hand side of the specified equations in the file *DE-modelequations.xls* are available in one of *CC* sheets of *HistoryData-EU27.xls*. Variables that do not fulfil this condition are listed in the error file;
- that variables used in the equations have values in the last observed year as specified in the *CountryTimeSet.xls* file. Variables that do not fulfil this condition are listed in the error file.

If all conditions are fulfilled, a *CONGRATULATIONS!!* message appears on the screen (Fig. 3.9). All specifications of *DE-ModelEquations.xls* are, given the satisfaction of all data checking requirements, automatically transferred into GAMS code and stored in *Equations_DE.gms*. GAMS code is not generated or updated if one or more of the data and variable conditions are not fulfilled. The file *ErrorOf_DE_Model.txt* contains a list of errors and suggestions as to how to solve identified problems.

Where *EViews* software has been used to estimate a country's AGMEMOD model, the AGMEMOD GAMS generator tool can be used to re-estimate the complete set of equations for that country. This facility can be useful after the updating of the AGMEMOD database with newly released annual data. The GAMS generator tool is activated by selecting

ModelSources \.1 Data and Model preparation phase by using Agmemod2Gams tool \ *1.1* Adjust CC_ModelEquation file and transfer equations into GAMS code

A screen appears, in which the country, for example *DE*, whose entire model is to be re-estimated can be chosen. By selecting the **Re-estimate all Equations in Eviews** option and by clicking on the **Start Processing** button, the GAMS generator tool starts the following procedure:

- it links the contents of *DE-ModelEquation.xls* with the contents of the Eviews file *DE-Eviews.wf1* and automatically re-estimates all equations of the German model;
- it checks the consistency between the mnemonics, data and equations used in *DE-Eviews.wf1*, *DE-ModelEquation.xls* and the data input files used;
- it generates updated GAMS code in *Equations_DE.gms*, based on the re-estimated set of equations.

3.3.4 Running the Model and Showing the Output

After the generation of the AGMEMOD computer version in GAMS, the model can be used to:

- conduct a baseline scenario for a country (group) or the whole EU;
- conduct alternative scenarios for a country (group) or the whole EU;
- look at the scenario results for a country (group) or the whole EU.

Before a selected scenario can be conducted, the active country set and the macroeconomic and policy assumptions underlying the chosen scenario must be defined. To define a scenario narrative, launch the menu option *Scenarios* from AGMEMOD user interface. A screen appears in which scenario(s) can be selected in the **Scenario to Run** box (Fig. 3.10). The three project information boxes are used to describe background information on the scenario(s), such as aims and narratives.

The buttons on the bottom right hand side of the screen are used to enter typical scenario narratives into the AGMEMOD model code. The first button, the **Edit Scenario Setting (gms)** button, opens the GAMS programs that control the (group of) countries that can be activated in the selected scenario (*SelectActiveCountriesInModel.gms*) and that define the specific macroeconomic and policy assumptions from the selected scenario in the file *AssumptionsInput.xls* (*SelectActiveSheetsForXlsScenarioInput.gms*). This button also opens the program

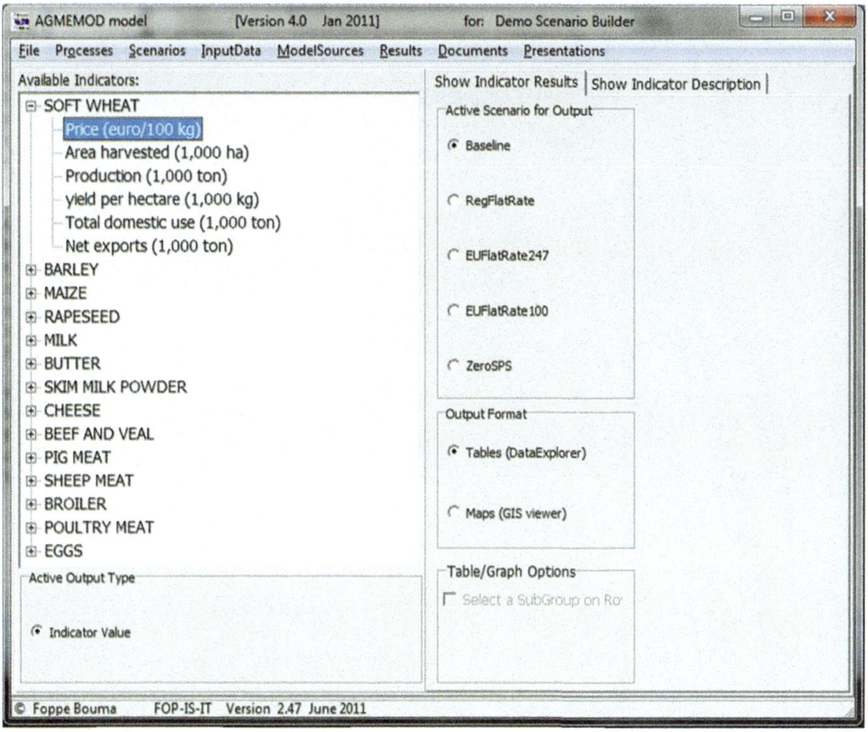

Fig. 3.10 Menu option that selects the scenario(s)

that determines the (group) of countries for which GAMS model output should be exported to MS-Excel files.

The second button, **Edit Scenario Xls data**, opens the *PolicyHarmon.xls* file that captures key policy harmonization data such as national budgetary ceilings, coupling rates, modulation rates, coupled payments etc up to 2020 for each country and for each scenario.

The third button, **Show Scenario RunInfo**, provides textual information concerning the last scenario run including countries in the selected country group and the policy assumptions used.

When the preferred country set as well as the appropriate macroeconomic and policy assumptions have been determined, the selected scenario, e.g. the baseline, can be generated by running the AGMEMOD model. Choosing the **Back to Processes** button activates the **Processes** screen that is composed of various process options. Which process option to select depends on whether data and/or equations have been revised in the preceding phases. The following three options are distinguished:

– **Read EU27 historical country data (same for all scenarios)**: should only be selected when data in *HistoryData-EU27.xls* have been revised since the importation of this large database is time consuming;

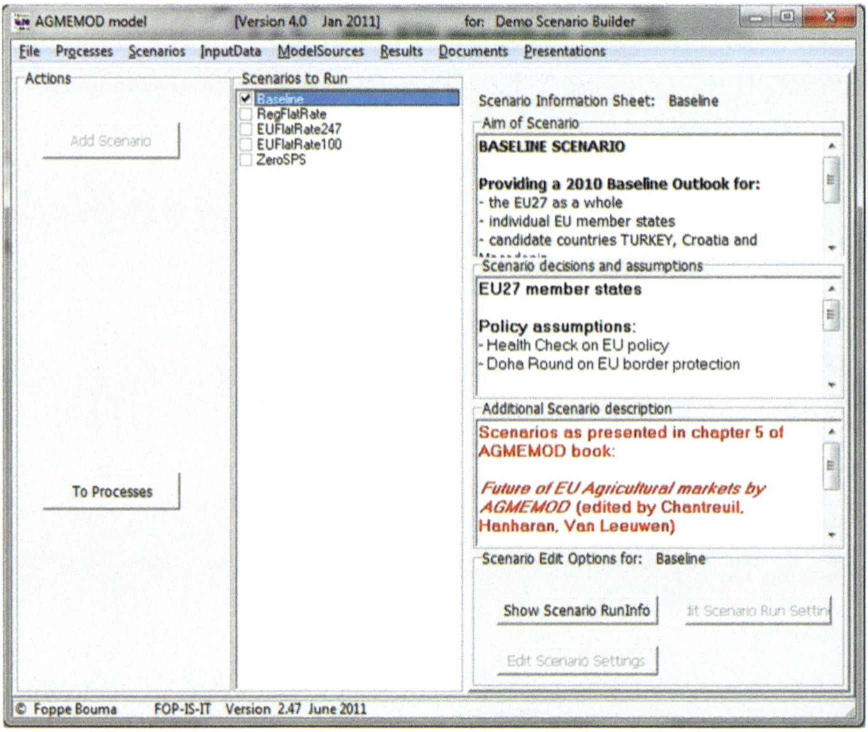

Fig. 3.11 Menu option that shows AGMEMOD scenario results

- **Read new Scenario data and run selected Scenarios**: reads in the exogenous scenario data from *AssumptionsInput.xls* and *PolicyHarmon.xls* and solves the model for the selected scenario(s);
- **Run selected scenarios (without reading in of Scenario data)**: runs the model for the selected scenario(s) without reading in the scenario data.

The last two options run and solve the AGMEMOD model for the selected scenario, which should only take a couple of minutes. A message appears on the screen when the solving process has been successfully finished. If for some reason a solution is not reached the file *Agmemod.lst* includes error messages that should aid in the resolution of the problem that gave rise to the failure to find a solution.

The **Processes** menu also contains possibilities to export scenario results to output files. The following three options are distinguished:

- **Create Excel output for selected Scenarios (CC-Results.xls files)**: this exports the output of the selected scenario(s), for the selected countries, to the MS-Excel file(s) *CC-Results.xls;*
- **Calculate scenario differences, Create Tables and Maps**: creates tables, graphs and maps that can be useful when analysing scenario results;
- **Apply Graph tool for current scenario**: to analyse and validate scenario outcomes.

To view scenario results select one of the following from the menu option *Results* of the AGMEMOD user interface:

Results \Tables (GDX data)
Results \Maps (GIS viewer)
Results \Tables (CC-Results.xls)

Depending on the choice of display media, scenario output is presented in tables, graphs and maps (Fig. 3.11).

3.4 Validation

The process of achieving plausible and useful modelling outcomes is a continuous one involving ongoing database improvement, model equation re-specification and re-estimation, analysis and discussion of model outcomes, etc. Simulation models, such as AGMEMOD, tend to change very rapidly over time due to emerging analytical demands and model and data update processes that can be ongoing simultaneously in the close to 30 AGMEMOD partners. The performance and quality of the combined model will depend importantly on the realization of a successful model validation process. Section 3.4.1 describes the stages of the AGMEMOD validation procedure, while Sect. 3.4.2 provides information on some helpful validation tools developed as part of the AGMEMOD projects ongoing research.

3.4.1 Procedure

The key aim of the country and EU versions of the AGMEMOD model is to generate projections for baseline and alternative policy scenarios. The quality of models that are mainly been designed for policy analysis is highly dependent on a successful validation process. Due to the importance of providing plausible and reliable baseline and scenario analysis, the validation of model outcomes is an integral part of the standard AGMEMOD model development and maintenance process. The model validation process is a key component of each baseline or scenario study and it always involves a sequence of revision rounds of the country models.

The main objective of the validation process is to improve the AGMEMOD model's capacity to generate plausible and sensible market outlooks and contribute impact assessment analysis of different policy options to the wider policy analysis and policy making community. The consistency of the models and their projections in agronomic and economic terms are checked and the models are improved if required. There are two phases in the validation procedure used in AGMEMOD studies.

In the first phase of updating the AGMEMOD model, an initial validation is carried out by the country teams (Erjavec et al. 2007). Partners check the consistency of the estimated behavioural production and consumption equations with theoretical requirements, necessary biological constraints, and standard statistical tests. This process usually takes place at the final stage of the estimation process (described in Chap. 2), during which in-sample and post-sample projections are used to support self-validation initiatives. Partners apply additional estimation and calibration work until the equations provide satisfying statistical results.

The second phase involves preparation of detailed country reports and templates to be used for external validation. Each validation report summarizes the model results, indicates the forces that drive the results and mentions the actions (if any) to be taken to improve the reliability of the model's outcomes. The country reports are distributed among partners and serve as basis for subsequent model improvements. This development work is the responsibility of the AGMEMOD partners.

Partners are encouraged to review and improve their model structure with support of agricultural market experts in their Member State. This interaction between partner and national market experts is a key element in the achievement of plausible baseline and scenario analysis of country level agricultural commodity markets.

3.4.2 Helpful Tools

The GAMS generator tool described in the previous section by providing a framework within which to build GAMS models has proved successful in reducing a number of problems encountered when working with a larger group of modellers. However, the tool by itself is unable to judge the suitability of a given estimated model specification with respect to the magnitude and sign of equation parameter values or a models capacity to project sensible and plausible commodity market projections. Tools with which to carry out within-sample analysis and/or shock analysis have been incorporated in the GAMS generator tool. Within-sample analysis and shock analysis can together with expert knowledge aid in the development and improvement of simulation models such as AGMEMOD model.

3.4.2.1 Within-Sample Analysis

The purpose of within-sample analysis is to examine the goodness-of-fit of estimated variables in comparison with their observed values over a pre-defined historical time period.

To conduct within-sample analysis at the country or the EU level, the MS-Excel file *CountryTimeSetInSample.xls* is used. Let *In Sample* be the scenario name for the within-sample analysis to be conducted and let the predictability of the 2000–2005 period by the Dutch model be the object of analysis. This analysis process is managed using the MS-Excel file *CountryTimeSetInSample.xls*, which can be viewed and edited by choosing the *InputData* menu option from the AGMEMOD user interface

InputData\AGMEMOD Excel input files\Time sets for In-Sample analysis-countries (*CountriesTImeSetInSample.xls*)

First, enter the "*NL*" row of the *ReferencePeriod* sheet and fill *ones* under the columns for years 2000–2005. Second, enter the '*NL*' row of the *RealSimulationPeriod* sheet, and include *ones* under the columns for years 2000–2005, but remove the *ones* under the columns for the remaining years. This modification ensures that the AGMEMOD model, when it runs the *InSample* scenario generates projections for the Dutch model over the 2000–2005 period which can then be compared with the observed values from the AGMEMOD database. The goodness-of-fit of the within-sample projections can be evaluated by comparing the results of both scenarios. A screen appears in which both scenarios must be selected in the **Scenario to Run** box. When the preferred country set (the Netherlands in this example) as well as the appropriate macroeconomic and policy assumptions have been determined (this must be exactly the same for both scenarios), the selected *Baseline* and *In-sample analysis* scenarios can be conducted by running the model. The **Back to Processes** button activates the **Processes** screen, and then select the option

– **Read new Scenario data and run selected Scenarios**

to read the exogenous scenario data from *AssumptionsInput.xls* and *PolicyHarmon.xls* and to solve the model for both scenarios. When the model has been solved successfully, the **Processes** menu contains possibilities to export scenario results into output files for further analysis (see Sect. 3.3.4).

3.4.2.2 Shock Analysis

The AGMEMOD framework includes a tool with which to conduct shock analysis and to calculate implicit model elasticities. This is useful when testing and assessing the performance and robustness of the country models. The impact of shocks to exogenous variables on endogenous variables of countries and the aggregate EU can be analysed. The following variables and parameter types can be changed:

- key prices (V parameter);
- world market prices (VWP parameter);
- macroeconomic variables (VMAC parameter);
- policy instruments (VPOL parameter).

The shock analysis tool is run by choosing the menu option *Scenarios* from the AGMEMOD user interface. A screen appears from which a scenario must be selected in the **Scenario to Run** box, for example the *Baseline*. The choice of which exogenous variable to be shocked is managed through the **Edit Scenario Setting (gms)** button on the bottom right hand side of the screen. This button opens the

GAMS program *SelectToDefineShocksOrElasticities.gms*. There are various options to choose from, such as

- calculate 'shock impact' (*YES* or *NO*)

  ► **$SetGlobal ShockCalculateShocks** *YES/NO*

- introduce an enduring 'one year' shock

  ► **$SetGlobal ShockParYear** *2009* {year 2009 is shocked}

- introduce an enduring 'long period' shock (*YES*) or not (*NO*)

  ► **$SetGlobal ShockAllYearsBeyond** *YES/NO*

- calculate point elasticities (*YES*) or not (*NO*)

  ► **$SetGlobal CalculateElasticities** *YES/NO*

When the preferred shock duration or elasticity calculation has been chosen, it can be applied to the selected *Baseline* scenario. The **Back to Processes** button activates the **Processes** screen, form there the option to select is

- **Read new Scenario data and run selected Scenarios.**

This reads the exogenous scenario data from *AssumptionsInput.xls* and *PolicyHarmon.xls* files and solves the AGMEMOD model. When the model has been solved successfully, the **Processes** menu contains an option to export scenario results to output files which are automatically given a 'shock-specific' name. The impact of the shock is calculated by comparing the results of the initial baseline scenario and the baseline scenario that includes a shocked variable or variables.

Box 3.1 contains an example of how to implement a shock to a key price. Shocks to endogenous variables must be treated differently from shocks to exogenous variables.

3.5 Conclusion

Simulation models such as AGMEMOD tend to change rapidly over time. New model versions and new scenarios are developed as the model is used in new and different research projects. This process could compromise the consistency between conceptual models and implemented computable models. This problem has been addressed by combining the knowledge of model building scientists and IT-scientists in the AGMEMOD project. In cooperation they have developed an AGMEMOD user interface with which to manage the building of the model's database and the country models, as well as the combination and the solving of these models and the presentation of scenario results. This computer framework has the following characteristics:

Box 3.1 Fifty Percent Shock of French Soft Wheat Price from 2009 Onwards

STEP 1 Select the *Baseline* scenario as reference situation in the **Scenario to Run** box of the AGMEMOD user interface

STEP 2 Select country (group) in *SelectActiveCountriesInModel.gms*

(a) The key price country (FR in this example) is shocked and must therefore be excluded from the EU country set: Select NO

► **$SetGlobal EU_TotalRun** *NO*

(b) Select the country (group) to be analysed (FR must be excluded)

► **$SetGlobal FR_InModel**
► **$SetGlobal NL_InModel** *NL*
► **$SetGlobal FI_InModel** *FI*
► **$SetGlobal IT_InModel** *IT*
....
etc.

STEP 3a Open *SelectToDefineShocksOrElasticities.gms*

(a) The model must be run for the selections made in steps 1 and 2. The shock effects are calculated by comparing the results of this Baseline scenario and the initial Baseline scenario (endogenous FR soft wheat key price)

► **$SetGlobal ShockCalculateShocks** *NO*

(b) Click on the **Back to Processes** button and select the option **Read new Scenario data and run selected Scenarios**

STEP 3b Open *SelectToDefineShocksOrElasticities.gms* again and define the key price shock

► **$SetGlobal CalculateElasticities**	*YES*
► **$SetGlobal ShockParName**	*V*
► **$SetGlobal ShockParElement**	*WS_PFN*
► **$SetGlobal ShockParCountry**	*FR*
► **$SetGlobal ShockParYear**	*2009*
► **$SetGlobal ShockAllYearsBeyond**	*YES*
► **$SetGlobal ShockEraseValuesBelow**	*0.01*
► **$SetGlobal ShockPercentage**	*50*

(a) click on the **Back to Processes** button and select the option **Read new Scenario data and run selected Scenarios**
(b) the shock impacts are automatically exported to ..**\CountryOutput** *Baseline_EU27_Shock_V_WS_PFN_FR_2009on.gdx*

- it integrates all Member State models into a combined EU version;
- it provides a transparent, flexible and consistent modelling tool, which is easily extendible with new commodities and countries;
- it makes baseline projections out to a 10 year horizon at the country and EU27 levels;
- it carries out and reports simulations and can help in the analysis of model results.

The development of the AGMEMOD user interface makes the AGMEMOD model more accessible for researchers and end-users and satisfies the flexibility condition that models must satisfy in order to meet requirements of various projects while also allowing for the maintenance of model version control. Finally, as the software is properly structured and documented, it can easily be passed on to other researchers or new team members.

References

Chantreuil F, Levert F (2007) AGMEMOD database documentation. Document number: AGMEMOD WP3 P1 D3. Project no. SSPE-CT-2005-021543

Cottrell A, Lucchetti R (2007) Gretl User's Guide. http://ricardo.ecn.wfu.edu/pub//gretl/manual/en/gretl-guide.pdf

Erjavec E, Regoršek D, Kavčič S (2007) Report on testing and validation of country models. Document number: AGMEMOD WP4 P21 D6. Project no. SSPE-CT-2005-021543

European Council (2003) Council Regulation (EC) R1782/2003

European Council (2009) Council Regulation (EC) No R73/2009

EViews (2011). http://www.eviews.com/

Food and Agricultural Policy Research Institute (FAPRI) (2010) FAPRI 2010 U.S. and World Agricultural Outlook. Available via http://www.fapri.iastate.edu/outlook/2010/. Cited 21 June 2011

Hanrahan KF (2001) EU GOLD model manual. Mimeo Rural Economy Research Centre, Teagasc. Dublin

Hanrahan KF, Levert F, Chantreuil F (2006) AGMEMOD Partnership. Notes and guidelines 3 "Mnemonic Protocol"

M'barek R, Bartova L (eds) (2007) GAMS simulation environment (GSE) – a user guide. EUR Number: 22547 EN. Available via http://www.jrc.es/publications. Cited 21 June 2011

McCarl BA (2010) McCarl GAMS user guide, Version 23.3. Available via http://www.gams.com/dd/docs/bigdocs/gams2002/mccarlgamsuserguide.pdf. Cited 21 June 2011

OECD-FAO (2010) Agricultural Outlook 2010. OECD, Food and Agriculture Organization of the United Nations. OECD Publishing, Paris

Van Leeuwen M, Tabeau A, Bouma F, Dol W (2008) Technical report on the combined model. Deliverable number: AGMEMOD WP6 P16 D8. Project no. SSPE-CT-2005-021543

Verhoog D (2000) Introduction to the agricultural information system of Eurostat. In: Pauli P (eds) Proceedings of the seventh IWG-AGRI seminar, Luxembourg, 5–7 July 2000

Part II

Chapter 4
EU Market Outlook

Trevor Donnellan, Frédéric Chantreuil, Emil Erjavec, Roberto Esposti, Kevin F. Hanrahan, Myrna van Leeuwen, Petra Salamon, and Guna Salputra

Abstract A key objective of AGMEMOD is to generate 10 years forward baseline projections for the main agricultural commodity markets of the EU and its Member States. This chapter provides detail on the AGMEMOD 2010 baseline projections. The baseline projections in this chapter are illustrative of the type of output produced by the AGMEMOD model that should be of interest to policy makers, market analysts and academics. Baseline results are provided for crops, livestock, dairy, fruits and vegetables.

Keywords Baseline scenario • Supply and use • Partial equilibrium sector modelling • AGMEMOD model • Use of the results

T. Donnellan (✉) • K.F. Hanrahan
Rural Economy Research Centre, Teagasc, Athenry, Co Galway, Ireland
e-mail: trevor.donnellan@teagasc.ie

F. Chantreuil
Université de Caen Basse Normandie UFR Sciences Economiques et Gestion, 17 rue Claude Bloch, BP 5186, F-14032 Caen cedex, France

E. Erjavec
Biotechnical Faculty, University of Ljubljana, Groblje 3, 1230 Domžale, Slovenia

R. Esposti
Department of Economics, Università Politecnica delle Marche, Ancona, Italy

M. van Leeuwen
LEI Wageningen UR, Alexanderveld 5, The Hague 2585 DB, The Netherlands

P. Salamon
Johann Heinrich von Thünen-Institute (vTI), Brunswick, Germany

G. Salputra
Latvia State Institute of Agrarian Economics, 14 Struktoru St, Riga, Latvia

F. Chantreuil et al. (eds.), *The Future of EU Agricultural Markets by AGMEMOD*, DOI 10.1007/978-94-007-2291-0_4, © Springer Science+Business Media B.V. 2012

A key objective of AGMEMOD is to generate 10 years forward baseline projections for the main agricultural commodities in the EU and its Member States. This chapter presents the baseline output produced by the AGMEMOD team using the model. The baseline projections highlight key medium term market developments. The baseline can also be used in conjunction with alternate scenarios to allow us draw some conclusions about future policy developments and their likely impact on EU agriculture.

Meaningful scenario analysis requires that a baseline is first developed to act as a benchmark for the assessment of any policy change. Chapter 5 examines various CAP reform scenarios and illustrates their economic impacts by comparing them with the projections made for the baseline outlook.

Section 4.1 presents the key assumptions underlying the baseline in respect of agricultural policy, the macroeconomy and world market commodity prices. Section 4.2 describes the EU market outlook to a 2020 time horizon, as generated by the AGMEMOD 4.0 version (2010).

4.1 Baseline Formulation

Macroeconomic indicators, world market prices and prevailing agricultural (trade) policies are key drivers behind the generation of a baseline outlook. These drivers are exogenous variables in AGMEMOD, meaning that they are determined outside the model. Under the baseline, the EU remains as currently structured, a 27 member union.

4.1.1 Marcoeconomy

Macroeconomic data are needed to generate baseline projections for the main agricultural commodities in the EU Member States. Historical data on macroeconomic variables such as population, inflation, per capita economic growth and currency exchange rates have been assembled at the country level. In general terms, macroeconomic projections used in the model were obtained from the national sources in the Member States. Figures 4.1–4.3 present the historic and projected values for population, GDP and inflation.

The projections indicate that there will be a decline in population in several eastern EU Member States, and a slight increase in population in western EU Member States, except in Germany. This can be largely attributed to westward migration by younger workers from Eastern EU Member States of the EU to western EU Member States.

The macroeconomic growth projections indicate that over the projection period there will be a slow recovery in growth following the financial crisis which began in 2008. Figure 4.2 shows that the rate of GDP growth in the EU15 and EU12 is negative in 2009 and 2010, while these growth rates are expected to become positive in 2011 and thereafter. As well as experiencing negative growth rates in the short term,

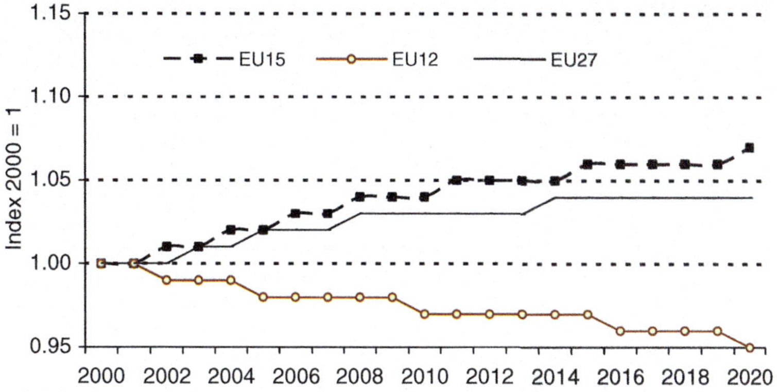

Fig. 4.1 Index of population for EU with projections to 2020 (Source: EU Member State national sources)

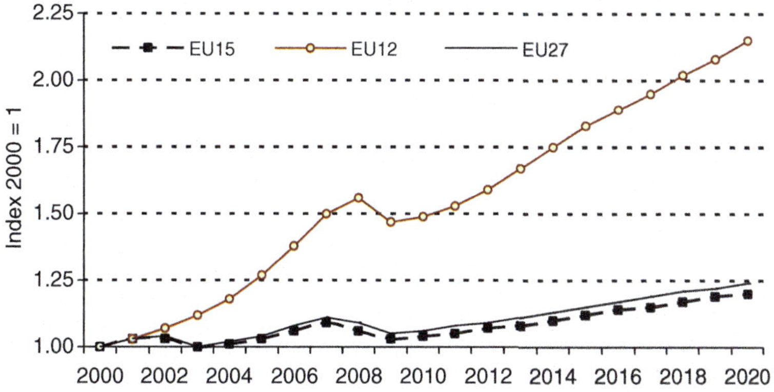

Fig. 4.2 Index of GDP for EU15, EU12 and EU27 with projections to 2020 (Source: EU Member State national sources)

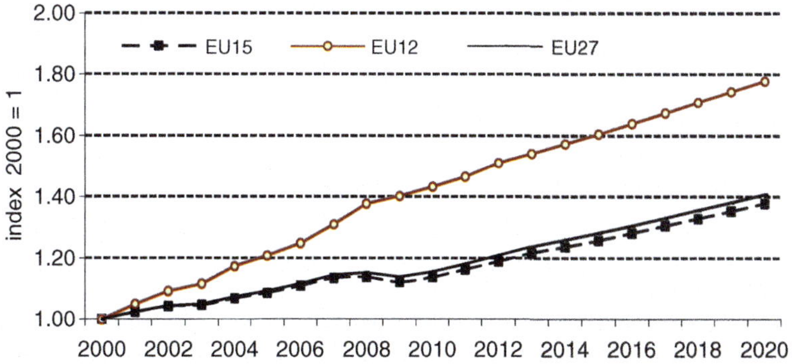

Fig. 4.3 Index of inflation for EU15, EU12 and EU27 with projections to 2020 (Source: EU Member State national sources)

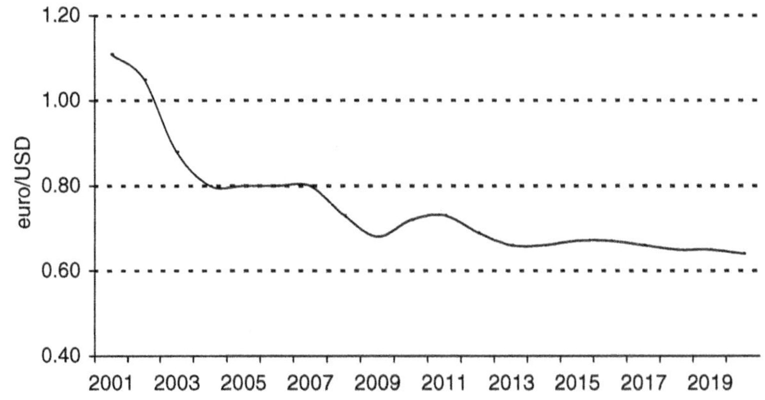

Fig. 4.4 Historic Euro/USD exchange rate with projections to 2020 (Source: FAPRI 2010)

Member States are also projected to face low or even negative rates of inflation in the early years of the projection period.

The exchange rate between the euro and the US dollar is a key macroeconomic variable, since it influences the domestic (euro) value of the exogenous US dollar world prices used in AGMEMOD. Projections for the euro/US dollar exchange rate (Fig. 4.4) are based on the observed exchange rate in 2010 and on the projected annual percentage change in that exchange rate in the period to 2020 which are published in the FAPRI World Outlook (2010). For the Eurozone countries, a weakening of the US dollar relative to the euro is projected over the next 10 years. For non-Eurozone countries, the exchange rate between national currencies and the US dollar is derived from projections of the exchange rate of the member states' currency with the euro.

4.1.2 World Market Prices

The world price projections used by the AGMEMOD model have been taken, in general, from the FAPRI World Outlook (2010). When necessary, e.g. for those commodities not considered by FAPRI, the FAPRI projections have been supplemented with projections from other sources such as the USDA Economic Research Service (ERS) and the OECD (2010). The world livestock (with the exception of lamb) and grain prices are US market prices. The world lamb prices used relate to New Zealand. Dairy commodity prices and oilseed, oilseed meal and oil prices are generally northern European export prices.

These world market prices are introduced in the EU key price equations and reflect the effects of global supply and demand on EU agricultural commodity markets. The developments of the world market price projections in the period to 2020 are presented in Figs. 4.5–4.8.

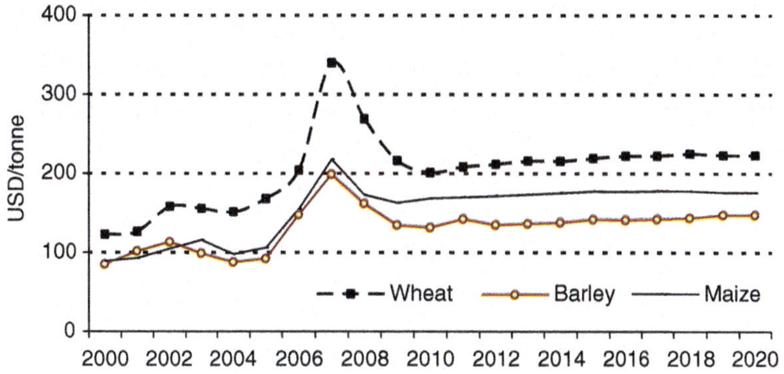

Fig. 4.5 World grain prices with projections to 2020 (Source: FAPRI 2010)

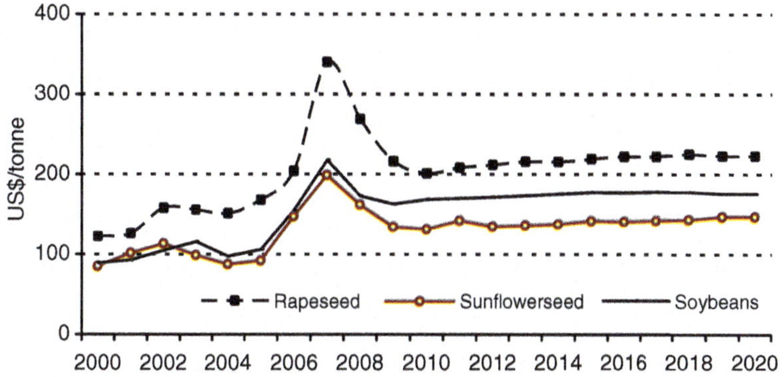

Fig. 4.6 World oilseeds prices with projections to 2020 (Source: FAPRI 2010)

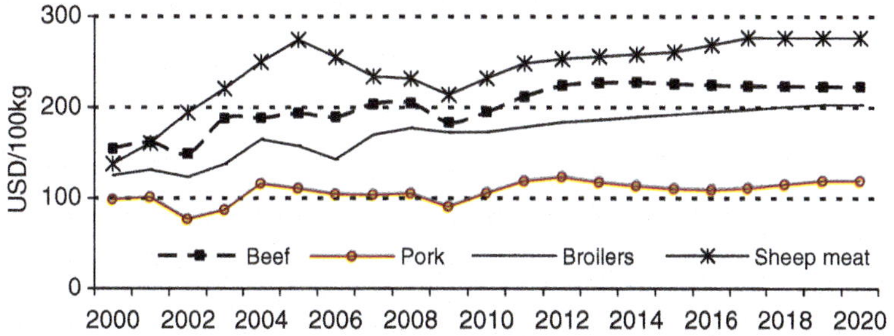

Fig. 4.7 World livestock product prices with projections to 2020 (Source: FAPRI (2010) and OECD-FAO (2010))

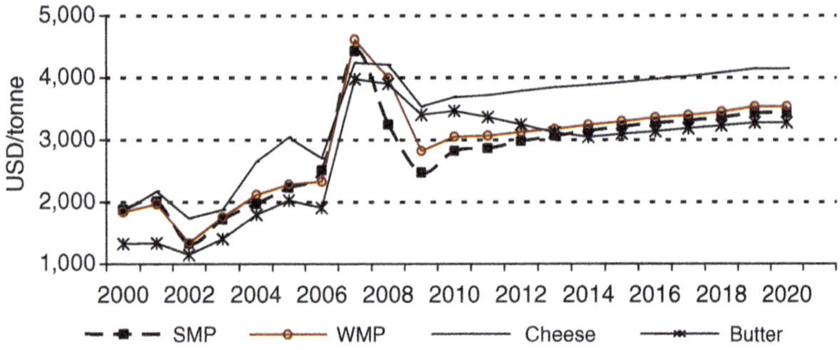

Fig. 4.8 World dairy products price with projections to 2020 (Source: FAPRI 2010)

Grains and Oilseeds: The general trend across all of the major agricultural commodities is one of rising world prices, which are driven by increasing demand around the world for grains and oilseeds. World prices for the major grains and oilseeds are projected to increase, particularly as competition for resources such as land from the biofuel sector pushes up food and feed prices in cereals markets and also because of the increasing derived demand for feed arising from the projected growth in global meat consumption.

Livestock: World beef, pork and broiler and sheep meat prices are also projected to increase over the projection period as income growth generates strong international demand. The rise in feed prices has knock on consequences as it increases the input prices for livestock and livestock product production, which impacts negatively on the projected production growth in these sectors and contributes to the projected increases in the world prices for livestock and livestock products (Fig. 4.7).

Dairy: World dairy product prices increased sharply in 2007 and 2008 due to a combination of factors. The growth in international demand for dairy products is quite strong and the availability of supplies to meet this demand has been affected by a gradual decline in the export surplus from the EU and a slowing in the growth of production in the southern hemisphere. World dairy commodity prices declined significantly in 2009, but increased again in 2010. It is projected that international dairy product prices will rise steadily over the duration of the projection period (Fig. 4.8).

It should be noted, however, that the above discussion relates to world prices as measured in US dollars. The very positive world price outlook across the main commodities is affected somewhat by the expected appreciation of the euro against the US dollar over the projection period. A consequence of this US dollar weakness is that the world price increases, measured in euro, are more moderate.

4.1.3 *Agricultural (Trade) and Biofuel Policies*

The first, key element of the baseline projections comprises the definition and implementation of the CAP at the EU country level. The baseline policy of the EU27 models reflects the 2008 CAP Health Check, which includes:

- the abolition of milk quota from 2015 with the agreed annual increases in milk quota, starting in 2009;
- the set-aside rate set to zero, starting in 2009;
- fully decoupled direct supports, starting in 2012;
- the application of Article 69;
- the continuation of (partly) coupled suckler cow and ewe premiums in those Member States that retained these payments;
- modulation rate rises to 14%, starting in 2012.

Based on the policy structure as described in section 2.4, a harmonized policy database at the EU country level has been generated and implemented in AGMEMOD. It incorporates all types of direct payments, reflects the necessary coherence between different types of EU CAP direct support elements and is sufficiently detailed to allow for impact analysis on the supply of agricultural products.

Second, with respect to agricultural trade policy developments, the baseline makes no assumptions concerning the outcome of the still ongoing Doha Development Round of the (WTO 2011). As no probable quantitative outcome to the negotiations was available when this baseline was produced, assumptions regarding the shape of a future agreement and the impact of the Doha Round on EU agriculture would be speculative.

Third, the baseline scenario takes account of the EU biofuel directive in the sense that 10% of the EU energy demand in transport must be met by biofuels by 2020. The PRIMES energy model (EC4MACS 2008) used by the European Commission (European Commission 2008) estimated that the energy demand in the EU27 in transport in 2020 will amount to 4,870 million gigajoules of gasoline (petrol) and 777 million gigajoules of diesel oil. Hence at the end of the decade, 5,650 million gigajoules of transport energy demand must be substituted by biofuels. AGMEMOD assumes that the division between bioethanol and biodiesel across the EU Member States will be equal, which generates a requirement for 20 million tonnes of rape oil and 85 million tonnes of cereals (soft wheat and maize). These bioethanol and biodiesel requirements have been exogenously calculated and comprise an additional category in the total domestic use of cereals and rape oil at EU level.

4.2 EU Baseline Market Outlook

The usual description of a baseline is a model based projection of the future, assuming that current policy remains unchanged over the projection period. The details of the narrative and assumptions underlying the baseline were presented in the previous

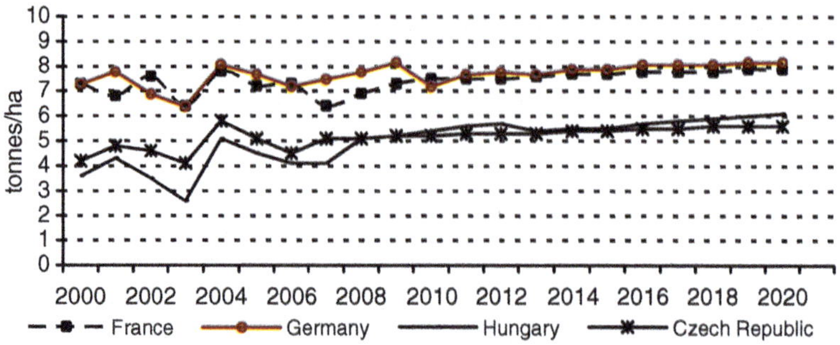

Fig. 4.9 Soft wheat yields in selected EU Member States and outlook to 2020 (Source: AGMEMOD 4.0 2010)

section. This section presents the EU baseline outlook for the period 2000–2020 which has been generated by the AGMEMOD 4.0 version (2010).

The results for the EU are presented at different levels of geographic aggregation: in the period 2000–2003 the EU consisted of 15 Member States (EU15), whereas it covered 25 Member States (EU25) in the period 2004–2006 and is represented as 27 states (EU27) over the period 2007–2020. For consistency and simplicity, the description of the baseline results will focus on the EU27.

4.2.1 EU Grains and Oilseeds Baseline

Under the baseline the total cereals area harvested in the EU is projected to decrease from 60.2 million hectares in 2007 to 57.8 million hectares by 2020. Across the cereal types, there is substitution in the area harvested away from barley, rye and oats and towards soft wheat and maize. Despite the projected reduction in the total cereals area harvested, total EU cereals production is projected to increase gradually as a result of the projected increase in productivity per hectare which more than offsets the reduction in area. Figure 4.9 shows the projected development of the baseline soft wheat yields per hectare for selected EU Member States.

Figures 4.10–4.12 present the medium-term baseline outlook for EU soft wheat, barley and maize markets. EU soft wheat and maize productions are projected to grow over the baseline projection period by about 1% per annum due to the higher prices arising from the assumed fulfilment of the EU biofuel directive. Barley production is not projected to increase in the projection period due to the removal of the barley intervention price in 2009.

The domestic use of cereals in the EU is expected to increase by 12% under the baseline from 2007 to 2020, but the pattern of increase differs significantly across the cereal types. The growth rates for domestic use of soft wheat and maize are projected to be highest, whereas the growth in domestic use of other cereals is below the average for cereals in aggregate. Furthermore, the EU domestic use of soft wheat

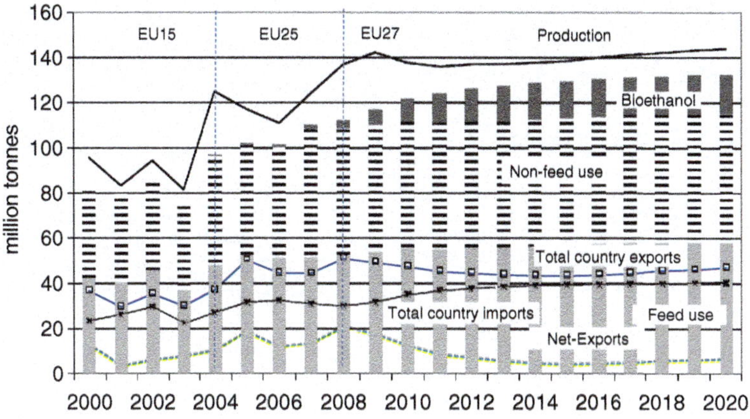

Fig. 4.10 EU soft wheat baseline outlook to 2020 (Source: AGMEMOD 4.0 2010)

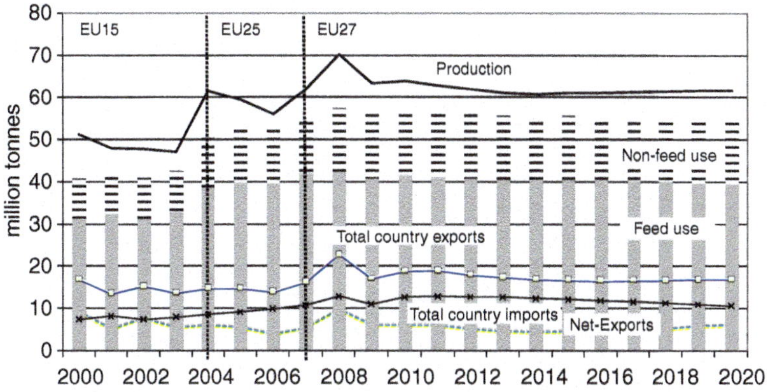

Fig. 4.11 EU barley baseline outlook to 2020 (Source: AGMEMOD 4.0 2010)

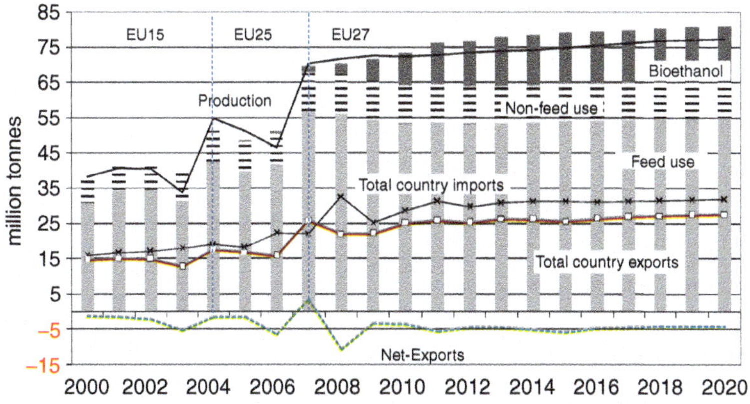

Fig. 4.12 EU maize baseline outlook to 2020 (Source: AGMEMOD 4.0 2010)

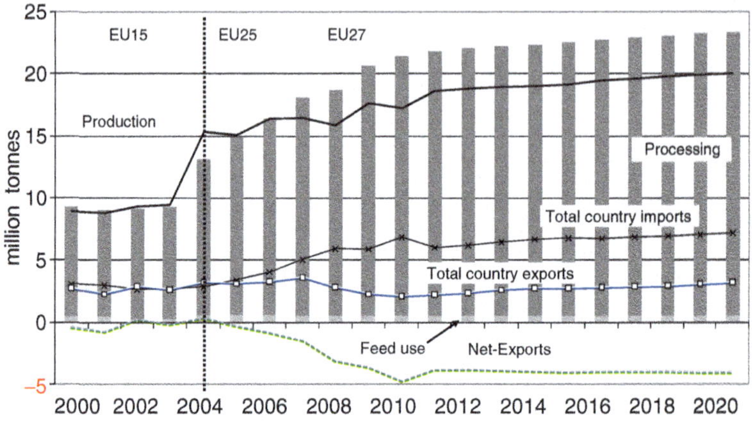

Fig. 4.13 EU rapeseed baseline outlook to 2020 (Source: AGMEMOD 4.0 2010)

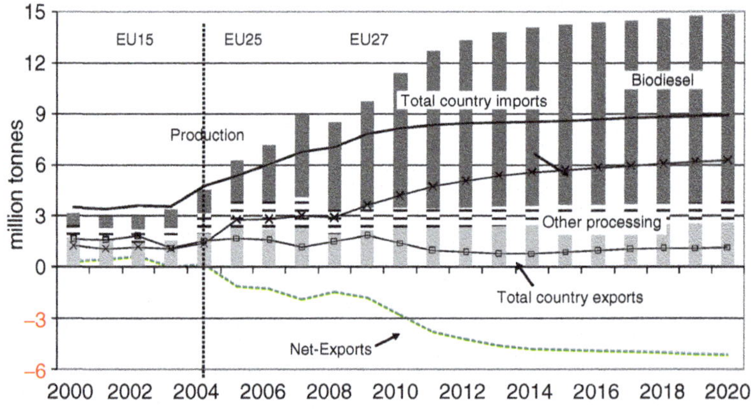

Fig. 4.14 EU rape oil baseline outlook to 2020 (Source: AGMEMOD 4.0 2010)

and maize is projected to grow at a slightly higher rate than EU production due to the assumed additional biofuel demand. It follows that EU net exports of soft wheat and maize are projected to decrease as increasing demand for bioethanol production leads to reduced availability of wheat supplies for export and increased net imports of maize.

Under the baseline the total oilseeds area harvested in the EU is projected to expand by 0.4% per year over the period 2007–2020. This strong demand growth is due to the higher demand for biodiesel. EU oilseeds production is projected to increase by 1.6% per year due largely to growth in yields.

Figures 4.13–4.15 present the medium-term baseline outlook for the rapeseed, rape oil and sunflower markets in the EU. In the projection period, the significant increase in the domestic use of rapeseed and rape oil leads to a growing EU net import position. To fulfil the EU biofuel directive, EU Member States are expected

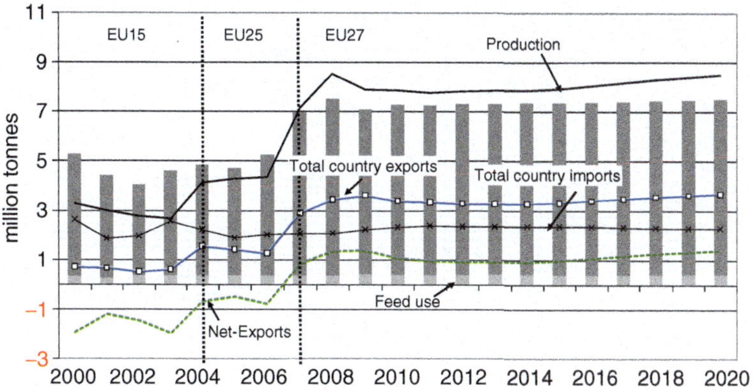

Fig. 4.15 EU sunflower baseline outlook to 2020 (Source: AGMEMOD 4.0 2010)

to become more dependent on rape oil imports from third countries or on substitutes for this oil. The consequential growth in the volume of rapeseed meal produced as a by-product of crushing the seeds for biodiesel feed stocks is projected to lead to lower animal feed meal prices in the EU.

4.2.2 EU Other Crops Baseline

This section examines the baseline outlook for the EU tomato, apple, olive oil and wine markets. Over the baseline projection period, prices for these crops are expected to remain stable and below their respective world market prices.

Tomatoes: EU tomato production is projected to expand due to a projected one percent annual growth rate in yields. The tomato area harvested is expected to grow slightly over the projection period, although the level reached by 2020 is significantly lower than its level over the period 1995–2005. This reduction in the harvested area has been the result of changes in policy intervention. Under the baseline, the growth in domestic use of tomatoes is driven by population growth (Fig. 4.16).

Apples: The baseline outlook for apples shows a similar evolution (Fig. 4.17) in the key variables to that seen in the case of tomatoes. EU apple production is projected to increase from 2007 to 2020, and this increase is entirely attributable to the projected increase in yields, since the area harvested is projected to decline. The projected decline in apple prices leads to an increase in projected total domestic use. Net exports of apples are projected to remain negative over the projection period.

Olive Oil: In the baseline the EU olive oil production is projected production increases over the period 2007–2020. The projected trend in strongly increase with the evolution observed over the years 2000–2007 (Fig. 4.18). The gradual decline

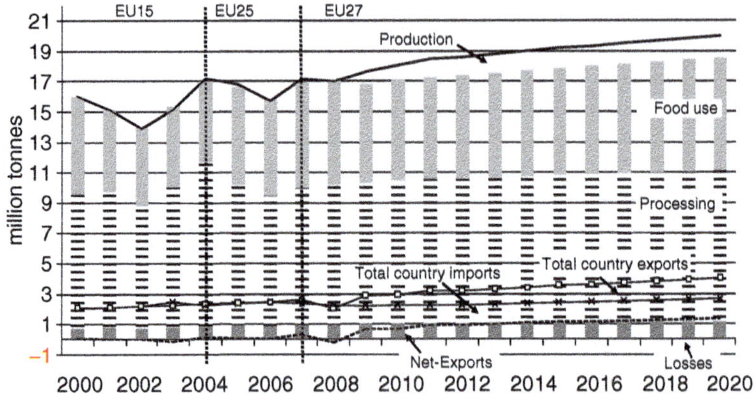

Fig. 4.16 EU tomato baseline outlook to 2020 (Source: AGMEMOD 4.0 2010)

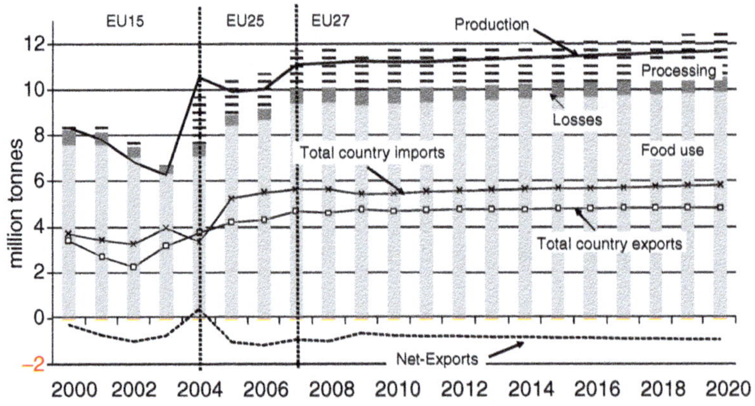

Fig. 4.17 EU apple baseline outlook to 2020 (Source: AGMEMOD 4.0 2010)

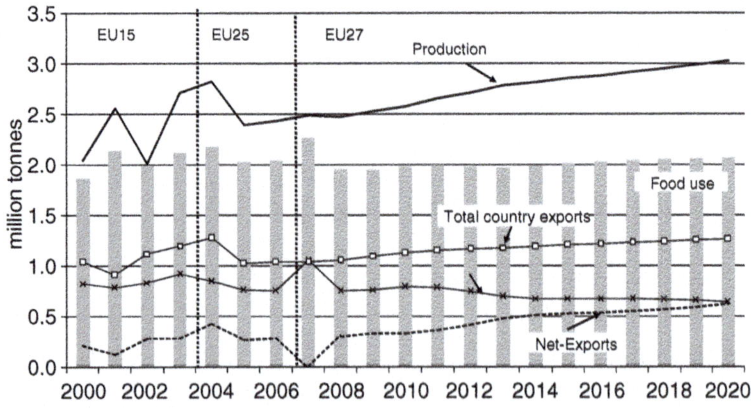

Fig. 4.18 EU olive oil baseline outlook to 2020 (Source: AGMEMOD 4.0 2010)

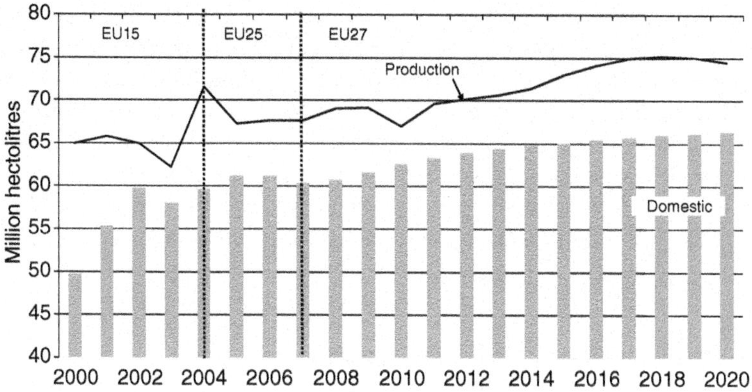

Fig. 4.19 EU quality wine baseline outlook to 2020 (Source: AGMEMOD 4.0 2010)

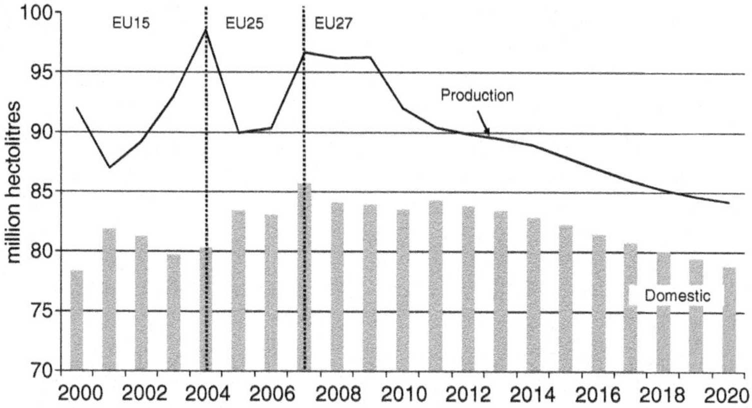

Fig. 4.20 EU table wine baseline outlook to 2020 (Source: AGMEMOD 4.0 2010)

in planted area over the projection period is offset by a 1.6% annual growth in yields per hectare. These developments are a reflection of ongoing long-term adjustments in the olive oil sector in major olive oil producer countries across the EU (notably Italy and Greece) where traditional extensive production is being progressively substituted by more intensive modern plantations. This intensification is supported by change in the nature of CAP support to olive oil producers which since 2008 is decoupled from production. Domestic use of olive oil is projected to decline over the period 2007–2020, but consumption in 2020 is still projected to be higher than in the year 2000. The cause of the projected decline in consumption is actually a move from lower to higher quality olive oil consumption. This change in the composition of the consumed product also occurs in the case of wine.

Wine: The baseline projections of the EU wine market reflect the more important changes the market will face, namely the new CAP regime and changing consumption behaviour (Figs. 4.19 and 4.20). On the supply side, it is projected that the wine area will decline only slightly. This is plausible considering the constraints the EU currently puts on the development of new vineyards and the allocation of area between quality and table wine. Under the baseline, production of quality wine is projected to increase, while table wine production is projected to decline over the projection period. Ultimately, total wine production is projected to remain more or less constant even though reported figures do not include production losses and other kinds of wine production. On the demand side consumers' preferences are projected to progressively shift towards quality wine, despite its higher price, in so doing substituting away from lower quality table wine.

4.2.3 EU Livestock and Meat Baseline

Under the baseline, no major policy changes that directly affect the livestock sector are assumed to occur. However, the implementation of policy changes such as the abolition of milk quota and the implementation of the Bioenergy Mandates will affect livestock markets through their impact on the number of animals available for slaughter and on the price of animal feedstuffs. Other factors such as the reduced rate of economic growth due to the ongoing recession and limited population growth in the EU also play a non-negligible role in determining the projected outcomes for EU livestock markets. In this subsection overall results for beef and veal, pig meat, broiler and lamb meat are presented.

Beef: Increases in feed prices and land scarcity, driven by the biofuel directives, in some Member States mean that meat prices are projected to increase over the baseline projection period. However, EU27 beef production is expected to decline by almost 5% to 7.9 million tonnes over the projection period, (most EU beef comes from the dairy herd which continues to decline), while EU27 beef consumption is projected to remain at around 9.1 million tonnes in 2020 (Fig. 4.21). The market for beef and veal continues to suffer from a negative consumer preference for red meat in a number of EU Member States. Under the baseline, the average amount of beef consumed per head in the EU is projected to be 18.4 kg by 2020. This level of consumption is comparable with that of 2002, but significantly below observed levels in the 1990s. Since 2004 the EU has been a net importer of beef and is projected to remain a net importer over the baseline projection period. The extent of the EU's net import requirement would increase if EU economic growth recovers more quickly and more strongly than anticipated in the current AGMEMOD baseline.

Pigs: With the higher prices for beef and lamb that are expected as a result of the (partial) decoupling of direct payments, the domestic uses of pig meat (Fig. 4.22) and poultry meat (Fig. 4.24) are projected to increase due to substitution effects. Average pork consumption per head in the EU is projected to reach 45.3 kg per year by 2020, a rise by 0.2% per annum compared to the level in 2007. Total domestic use of pork is projected to increase by 3.7% over the same time period.

Particularly low pig meat prices in recent years, especially relative to cereal (feed) prices, has led to an exit of producers from pig meat production. This pattern

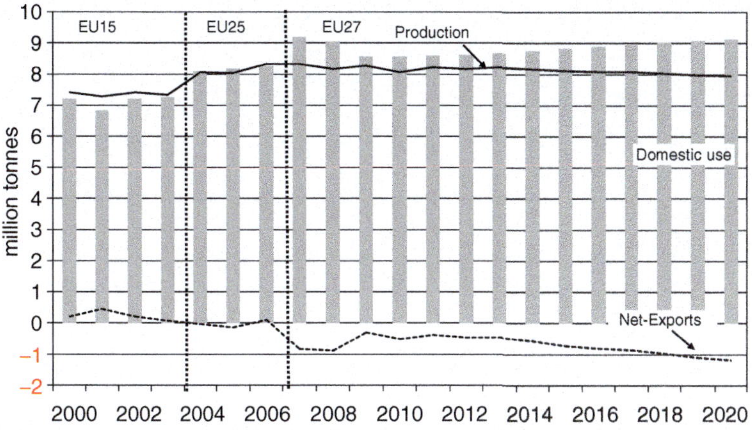

Fig. 4.21 EU beef and veal baseline outlook to 2020 (Source: AGMEMOD 4.0 2010)

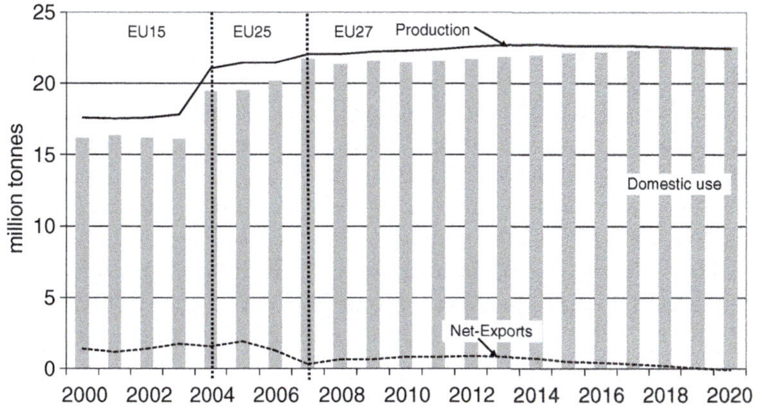

Fig. 4.22 EU pig meat baseline outlook to 2020 (Source: AGMEMOD 4.0 2010)

has been reinforced by EU and national environmental policies which have had an adverse impact on production costs. No recovery in pig meat price is foreseen, leading to a further decline of 6% in sow ending stocks over the period 2007–2020. However, more efficient pig production is projected to result in higher slaughter weights and overall EU pig meat production is projected increase by almost 2% over the projection period.

Sheep: Figure 4.23 shows that, under the baseline, the total domestic use of sheep meat in the EU will increase slightly over the period 2007–2020. This increase in domestic sheep meat use is driven by population growth, as the average consumption per head is projected to remain stable under the baseline.

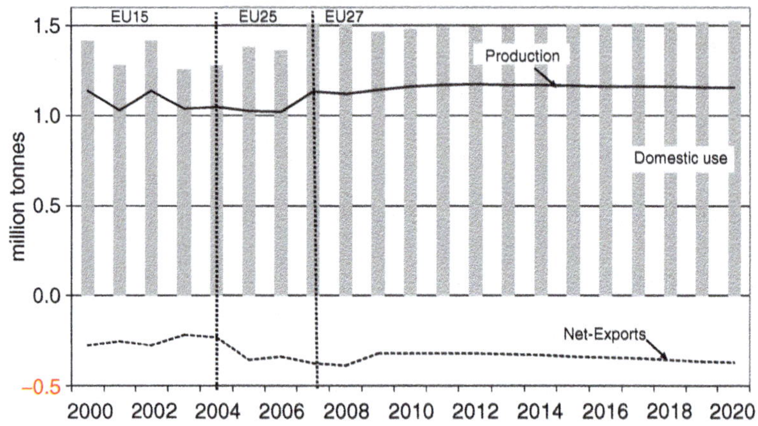

Fig. 4.23 EU sheep meat baseline outlook to 2020 (Source: AGMEMOD 4.0 2010)

The EU sheep meat market is protected by high tariffs and the EU imports lamb from New Zealand under a tariff rate quota. The overall price story for the EU is expected to be stable and largely determined by the market situation in France, the UK and Ireland. The projected growth in EU sheep meat production will be limited to 0.1% per annum over the period 2007–2020.

EU sheep meat price levels are close to the world market price levels and remain strongly driven by world market conditions and exchange rates. However, price differentials for different types of sheep meat (light lamb in southern Europe and heavy lamb in northern Europe) are projected to persist. The baseline projection to 2020, suggests that the EU as a whole will remain a strong net importer of sheep meat.

Poultry: Under the baseline, the situation in the poultry sector is characterised by a marked increase in the projected consumption per head of 0.9% per annum. EU average consumption reaches 26 kg per capita by 2020. Driven by projected poultry price increases and continuing strong technology incentives, EU poultry production is projected to increase by 16% over the period 2007–2020 (Fig. 4.24). A key element of this production growth is projected to occur in Poland.

4.2.4 EU Milk and Dairy Products Baseline

In the past, EU domestic dairy commodity markets were insulated from lower world market prices by EU policy in the form of intervention (reference) prices, import tariffs and export refunds. Increasingly, world prices are now playing a more important role in determining EU dairy product prices.

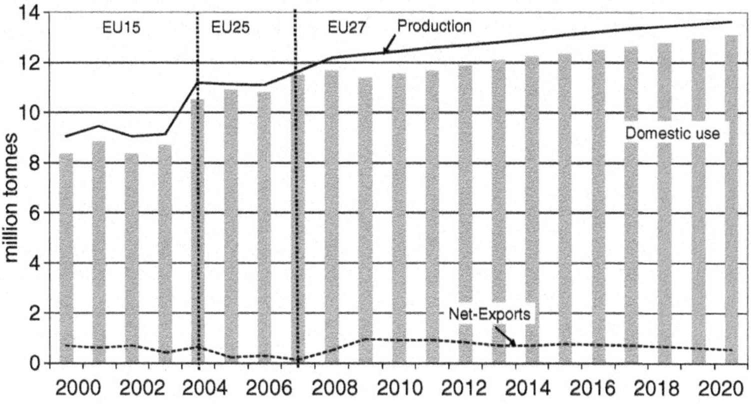

Fig. 4.24 EI poultry meat baseline outlook to 2020 (Source: AGMEMOD 4.0 2010)

Cuts in intervention prices and the suspension of export refunds, in combination with limited growth in EU domestic dairy product consumption, has reduced the spread between the world and EU market prices, and have consequently reduced the degree to which border policies insulate EU dairy markets completely from price fluctuations originating at the world market level.

Reflecting these developments prices for dairy commodities in the AGMEMOD model are driven by self-sufficiency ratios, policy measures such as intervention (reference) prices, world market prices and trade measures.

In the future EU dairy price stabilization measures are only likely to be considered when domestic prices fall below support prices. However, prices on the world market are volatile, and are likely to vary considerably around projected annual averages due to economic considerations on the demand side and due to yield variations associated with weather patterns on the supply side. The world market for dairy products is small relative to global dairy production/consumption which makes world dairy commodity markets particularly susceptible to even small variations in global production and consumption. As movements of the exchange rate between the euro and the US dollar normally add to price fluctuations, the future exchange rates will also play an important role in future price volatility at EU level.

Assuming normal weather conditions, it can be expected that world dairy product prices will remain lower than the, exceptional, levels achieved in 2007/2008. The baseline assumes the removal of the quota system in 2015 as established in the CAP Health Check agreement. By 2020 total EU milk production is projected to be 4.8% higher than in 2007. Yields per cow are projected to increase over the baseline at an annual rate of 1.3%. Given increased yields, the stock of dairy cows in the EU27 may still decline significantly without affecting milk production, which will have a negative impact on EU beef production, which is mainly sourced from the EU dairy herd.

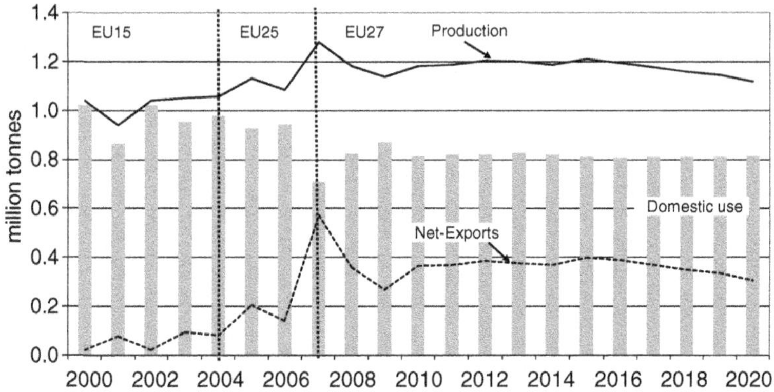

Fig. 4.25 EU skim milk powder baseline outlook to 2020 (Source: AGMEMOD 4.0 2010)

EU Dairy Product Mix: It is projected that there will be some reorientation of the EU dairy product mix in the sense that cheese production is expected to increase and the production of intervention products is projected to decrease. However, the tendency towards reduced production of intervention products, notably skim milk powder, will be tempered by the impact of higher world market prices. The rise in EU cheese production will broadly follow the projected increase in EU cheese consumption, while the reduction in butter and skim milk powder production reflects, in principle, the increased possibilities for alternative uses for milk.

SMP: In the case of skim milk powder (SMP), international prices maintain EU production at a relatively high level (Fig. 4.25), despite the fact that the protein demand for cheese manufacturing, supplemented by the demand for other fresh products, leads to an increased demand for milk protein within the EU. Under the baseline the domestic use of skim milk powder is projected to decline slightly, which reflects a projected drop in the feed use of skim milk powder and higher product prices. Baseline EU SMP net exports are projected to follow a similar path to that of SMP production.

Butter: Although the baseline world butter price is lower than the EU intervention price level, the projected gap between EU domestic and world market price levels will be reduced. This is due to the institutional price cuts within the EU and increased worldwide demand accentuated by reduced butter exports from the EU due to the lack of subsidies required to make third country exports competitive. EU butter production is projected to remain stable, whereas its net export position is increasing (Fig. 4.26).

Cheese: Baseline EU cheese prices are projected to be a bit closer to the world market price, but a gap between both prices is still projected to persist. This gap is unlikely to be bridged as prices also reflect perceived quality differences. Despite the

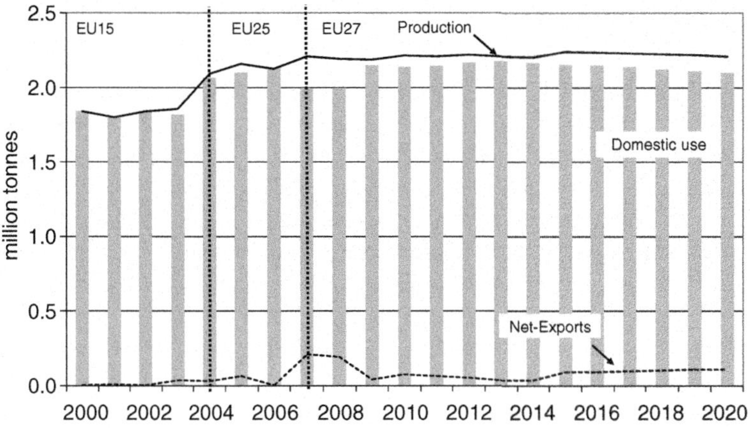

Fig. 4.26 EU butter baseline outlook to 2020 (Source: AGMEMOD 4.0 2010)

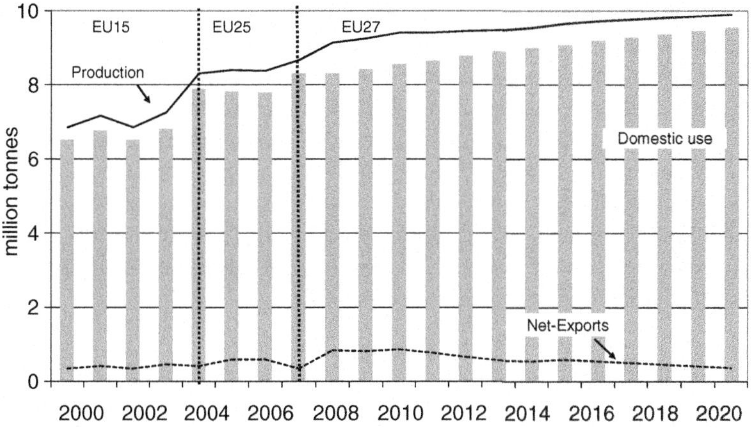

Fig. 4.27 EU cheese baseline outlook to 2020 (Source: AGMEMOD 4.0 2010)

price gap, quality differentials allow considerable EU exports without export subsidies to take place. Relative price changes between cheese and other dairy products as well as the lowering of intervention prices, particularly that of butter, is projected to result in the use of more milk in cheese production and less in intervention products. Thus, cheese production is expected to increase by 1% per year under the baseline (Fig. 4.27), which in turn limits the prospects for any cheese price increase over the baseline projection period. The rate of growth in cheese consumption is projected to exceed the rate of growth in cheese production, with the result that net exports of cheese from the EU are projected to decrease.

4.3 Conclusion

The motivation behind the development of a baseline scenario is its usefulness as a yardstick against which to measure the impact of possible or actual policy changes. The baseline is developed using policy, macroeconomic and population data from sources exogenous to the AGMEMOD model. In the baseline, rising US dollar world prices are observed for most commodities, but the strength of the euro over the projection period means that the increase in prices in euro terms is more moderate.

At EU level there is a modest change in total cereal production with rising yields being offset to some degree by a declining area under cereals. By contrast, the production of oilseeds increases in the EU due to an increase in both the yield and area harvested. EU domestic consumption of cereals and oilseeds increases at a greater rate than the increase in production, with the result that the EU's net export position deteriorates. Strong growth in EU olive oil and quality wine production takes place and a decrease in EU domestic consumption allows for an increase in the exports of both commodities.

In the meat and dairy sectors, growth in EU domestic use outstrips production with the result that the EU net export position deteriorates.

Overall, the baseline scenario results show that the reform recently agreed (the CAP Health Check agreement of November 20, 2008) has a pretty limited impact. This is not a criticism of the reform agreement since from the outset the intention of the reform was to consolidate the changes made in the 2003 CAP reform.

The phasing out of the EU milk quota represents the principal agricultural policy change across all commodity categories. The analysis presented in this chapter suggests that the increase in the rate of modulation agreed by the Council will have only very modest impacts on agricultural production. The above results also suggest that the freedom to retain coupled suckler cow and ewe premium limits the scale of any adjustment in the cattle and sheep sectors.

References

EC4MACS (2008) The PRIMES model. Interim report on modelling technology. Available via http://www.ec4macs.eu/home/reports/Interim%20Methodology%20Reports/7_Capri_interim%20_methodology300408.pdf. Cited 21 June 2011

European Commission (2008) Health check of the CAP (current situation, Commission proposal and Council outcome). Available via http://ec.europa.eu/agriculture/healthcheck/index_en.htm. Cited 21 June 2011

FAPRI (2010) U.S. and World agricultural outlook. FAPRI staff report 10-FSR 1. Available via http://www.fapri.iastate.edu/outlook/2010/. Cited June 21 2011

OECD-FAO (2010) Agricultural outlook 2010. OECD Publishing, Paris

WTO (2011) Doha development agenda: negotiations, implementation and development. Available via www.wto.int/english/tratop_e/dda_e/dda_e.htm. Cited 21 June 2011

Chapter 5
The CAP Beyond 2013

Emil Erjavec, Frédéric Chantreuil, Trevor Donnellan, Roberto Esposti, Kevin F. Hanrahan, Myrna van Leeuwen, Petra Salamon, and Guna Salputra

Abstract The main purpose of the chapter is to demonstrate a policy analysis application of the AGMEMOD model and to provide an illustration of the range of policy scenarios that can be analysed using the AGMEMOD model. Specific results are provided for a scenario which looks at the impact of equalizing the level of direct payments per hectare across the EU. We thus present, in this chapter, the results for different alternate policy scenarios along the EU and individual Member States dimensions, illustrating the appropriateness of the policy harmonization approach for such analysis.

Keywords Common Agricultural Policy • EU Budget • Regional flat rate • EU-wide flat rate • SPS abolishment • Partial equilibrium sector modelling • Agmemod model

E. Erjavec (✉)
Biotechnical Faculty, University of Ljubljana, Groblje 3, 1230 Domžale, Slovenia
e-mail: emil.erjavec@bf.uni-lj.si

F. Chantreuil
Université de Caen Basse Normandie, UFR Sciences Economiques et Gestion, 17 rue Claude Bloch, BP 5186, F-14032 Caen cedex, France

T. Donnellan • K.F. Hanrahan
Rural Economy Research Centre, Teagasc, Athenry, Co Galway, Ireland

R. Esposti
Department of Economics, Università Politecnica delle Marche, Ancona, Italy

M. van Leeuwen
LEI Wageningen UR, Alexanderveld 5, The Hague 2585 DB, The Netherlands

P. Salamon
Johann Heinrich von Thünen-Institute (vTI), Brunswick, Germany

G. Salputra
Latvia State Institute of Agrarian Economics, 14 Struktoru St, Riga, Latvia

F. Chantreuil et al. (eds.), *The Future of EU Agricultural Markets by AGMEMOD*,
DOI 10.1007/978-94-007-2291-0_5, © Springer Science+Business Media B.V. 2012

During the period 2010–2012, the direct payments systems of the CAP will be intensively discussed as part of the broader CAP reform process that will set the policy framework for EU agricultural policy for the period 2014–2020. The decisions on reforming the CAP direct payment systems will also depend on the post 2013 EU budget debate outcome. Any prediction as to the nature of the long-term changes to the CAP is speculative, with changes likely to depend largely on the division of power between the reformist and more conservative Member States (Garzon 2006; Swinnen 2008) and possibly external factors such as progress in the WTO negotiations (Daugbjerg and Swinbank 2007).

Esposti (2008) in a review of recent model based economic analysis of the CAP concluded that the long-term perspective of CAP after 2013 at the EU27 level, in the dimensions being currently publicly discussed (Buckwell 2007; Swinnen 2008; Begg et al. 2008; ECORYS 2008) and even the current shape of CAP under its different schemes, is not completely and systematically covered in any model or model conglomerates application. Esposti argues that this is because existing modelling approaches still do not correctly represent some important details of how agricultural policy is implemented at the Member State level. These deficiencies are largely due to the Member State heterogeneity and increasing complexity of the CAP as well as the nature of the policy questions posed by the possible future development of the CAP.

In this chapter, four possible post 2013 CAP reform scenario are outlined and their impact on the EU agricultural markets is analysed using the AGMEMOD combined model of European agriculture. The post 2013 CAP reform scenarios analysed reflect some of the content of recent public debates concerning the long term future of the CAP. The main hypothesis is that the implementation of such reform scenarios could lead to significant changes in the budgetary distribution of direct payments between Member States, whereas the impact on EU27 agricultural commodity production, while negative, would be relatively minor.

The chapter is structured as follows. Section 5.1 outlines the perceived difficulties with the current CAP Pillar I direct income support policy. Section 5.2 sets out the CAP reform scenarios evaluated. Section 5.3 details the results of the analysis conducted at both the EU 27 and Member State levels, while Sect. 5.4 discusses the results and draws some conclusions.

5.1 CAP Direct Payment Issue

The Fischler reform in 2003 changed the form of CAP direct income support payments by introducing decoupled direct income supports, though it largely preserved the scope and distribution of funds across Member States and types of agricultural holdings (Swinnen 2008). Policy modifications under the CAP Health Check agreement of 2008 followed the direction established in 2003 by further decoupling direct payments, increasing the rate at which payments are modulated and allowing Member States to switch from historical to regional flat area payment regimes. Despite the almost continuous reform process the pressure for further CAP reform has not abated, with the on-going EU Budget Review and parallel CAP reform process almost certain to lead to further agricultural policy change in the medium term.

The accession of new Member States in 2004 and 2007, when combined with the 2003 Fischler reforms, introduced a large degree of agricultural policy heterogeneity to the CAP by comparison to the Agenda 2000 policy framework. Within the parameters of the 2003 Fischler reform (and the subsequent CAP Health Check agreement) EU Member States have some flexibility in the degree to which they must decouple direct payments and in the choice of payment model used to implement the decoupled Single Payment Scheme (SPS). The EU15 Member States can implement the SPS by either granting historical support level to farmers or by using a regional flat area payment direct income support scheme. EU15 Member States also have some flexibility in the degree to which the link between production and receipt of direct income support is retained. New Member States are still allowed to use the transitional Single Area Payment Scheme (SAPS) support system. One of the advantages of the SAPS is the flexibility that it provides to Member States (during the transition period) to provide additional national funding to agriculture from national budgets in coupled and decoupled forms. EU funded support within the SAPS system must be totally decoupled from agricultural production.

Under the provisions of the CAP Health Check agreement of 2008 the diverse agricultural policy systems permissible under the CAP may be gradually equalized over the period 2010–2013 through the decoupling of direct payments that under the Fischler reforms could be retained by Member States as coupled direct payments, and via the voluntary switch from historical SP schemes to regional flat area SP scheme in old Member States that currently use historical payment models.

Contributions by Member State to the EU budget for the period 2014–2020 are unlikely to increase significantly, while pressure from net contributor Member States to reduce CAP spending is set to increase (Begg 2005; Begg and Heinemann 2006; Begg et al. 2008; ECORYS 2008). There is also a realistic possibility of a re-nationalization of the Pillar I of the CAP, i.e. that all Member States will be required to co-finance supports from national funds, a provision which has been publicly discussed by the EU budget Commissioner (Grybauskaitė 2008).

Pressure for greater uniformity in the level of direct payments across Member States will increase. In addition, average payment amounts will probably decrease due to the pressure from some Member States for their abolition, as they account for two thirds of the CAP budget. The continued existence of direct payments may hinge on reducing their redistributive nature (Begg and Heinemann 2006; Cipriani 2007) and on the search for a new rationale for their existence, such as ensuring public goods provision by agriculture (OECD 2003; Yrjölä and Kola 2004; Begg et al. 2008; Bureau and Mahé 2008).

In the context of the on-going and parallel EU Budget Review and CAP reform negotiations the following changes to direct agricultural payments could be considered:

- a reduction in the national envelope for direct payments. This solution, while perhaps politically realistic, retains the main negative distributional effects of the CAP and does not provide a justification for the continued existence of the CAP;

- the introduction of a new form of direct payment supports, such as an EU wide flat area payment or other more regionally uniform types of payments, could make the CAP more targeted in term of payments for non-commodity outputs related to agriculture;
- the use of modulation "savings" for other purposes outside of Pillar I of the CAP. This outcome would retain the unequal distribution between Member States of CAP spending and would therefore be likely to be controversial.

Reductions in Pillar I CAP funding from the EU budget could also be achieved if these payments were co-financed from national budgets (i.e. via a re-nationalization of Pillar I of the CAP). Different levels of co-financing could also address the problem of over-paying for the supply of public goods in different Member States, given that these goods are unlikely to be valued to the same extent throughout the EU. Incentivising voluntary co-financing and the prevention of policy inconsistency could be achieved by setting upper and lower limits for co-financing within Pillar I.

The effects of different potential changes to direct payment regimes depend on the direct payment scheme models used under the current policy dispensation in 2013. EU Member States can be broken into two groups on the basis of the type of direct payment scheme they are likely to use in 2013. The first group will in 2013 be utilizing an historical payment model, with the majority of these Member States also retaining at least some coupled measures. Countries in this group include Austria, Belgium, Greece, France, Ireland, Italy, Netherland, Portugal and Spain.

Member States that operate so-called hybrid direct payment schemes that retain at least some historical basis for the support farmers receive per hectare could also be included in the historical payment model group, this group includes Sweden, Slovenia and Luxembourg. The impact of a move to a flat area payments basis for direct payments would be expected to affect the agricultural sectors of the Member States in this group with particularly important effects likely to arise where Member States have retained coupled direct payments. Most of these coupled payments are linked to livestock production and the negative impacts of policy changes on these sectors are likely to be largest.

The second group of Member States that can be distinguished on the basis of the direct income support payment system used are those that in 2013 will use different types of flat area payment systems. In this group are EU15 Member States such as Denmark, Finland, Germany, Malta and the UK. The 10 new Member States (with exception of Slovenia and Malta) will operate the SAPS in 2013 which is based on uniform values of payment per hectare. Other things equal, the effect on this group's agricultural production and agricultural incomes of changes to direct payment policies that involve movement to flatter area payment systems should be smaller when compared with the impact on the first group. This is because these countries by 2013 will already operate a flat area payments model and in general will have retained fewer production linked direct payments.

5.2 Methodological Concept and Scenarios Definition

In the AGMEMOD combined model, decoupled payments have supply inducing impacts. The supply inducing impact of decoupled payments is differentiated on the basis of whether or not the payment is paid on an historical basis or on a regional (flat area) payment basis. The supply inducing impact of a euro of production decoupled support that is paid on an historical basis is assumed to be greater than the supply inducing impact of a euro of decoupled support that is dispensed on a regional payments (flat area payment) basis (Salputra et al. 2011).

The assumption that payments that are decoupled from production have at least some supply inducing effects is widely used in the partial equilibrium policy modelling literature. Economic theory suggests that lump-sum payments have no effect on production when markets are complete. However, under imperfect labour, credit, or insurance markets, decoupled payments could influence supply (Burfisher and Hopkins 2003; Chau and de Gorter 2005; Hennessy 1998; Roe et al. 2003). A review by the OECD (2001) suggests that, in addition to the effect which imperfect markets have on the production impact of decoupled payments, decoupled payments that are associated with restrictions on the use of land, cross compliance conditions and that create expectations concerning entitlement to future payments, can affect the degree to which so-called production decoupled direct payments influence production.

Relative to the baseline, in all of the four scenarios analysed, direct payments that were coupled to production under the baseline were fully decoupled. As a result the *reaction prices* (see Chap. 2) that capture the supply inducing impact of coupled and decoupled direct payments change relative to the baseline in most Member States. Under the reform scenarios analysed, the reductions in reaction prices in some Member States are in part due to the extension of decoupling agreed under the CAP Health Check. In addition, reaction prices are reduced relative to the baseline as the model of the SPS implementation changes under the scenarios analysed. Movement from an historical model in some Member States to a national or EU wide flat area payment model generally leads to lower reaction prices. Significantly, in those Member States where flat area payments per hectare in 2013 are lower than the average payment in the EU27, the move to an EU wide flat area payment leads to an increase in reaction prices.

5.2.1 Baseline Scenario

This scenario, described in detail in Chap. 4, involves the continuation of European agricultural policy as agreed in the CAP Health Check. Under the baseline, the mix of historic, regional, and dynamic hybrid direct aid schemes with coupled payments (where EU Member States have chosen to retain them) continues along with the mandatory elements of the Health Check decisions implemented through to the end of the projection period in 2020. Rates of direct payment modulation are increased, milk quota and set aside are abolished, and direct supports related to production are to be fully decoupled with the exception of some beef and sheep payments. The CAP budget national envelopes remain at their currently agreed levels.

5.2.2 Regional Flat Rate Scenario

Under the Regional Flat Rate (RFR) scenario all Member States currently using a historical payment model or hybrid model with a historical component move to a regional flat area payment model from 2014 onwards through a series of three annual changes in the value of existing entitlements. The annual reductions in the value of per hectare entitlements can be no more than 50% of the difference between the final value of the entitlement on completion of the move to a flat rate payment model and the value of the entitlement under the baseline.

Because the AGMEMOD model is based on models that are constructed at a Member State level of aggregation it is not possible to model sub-national regional flat area payments. In implementing the Regional Flat Rate scenario the regions analysed are taken to be equivalent to nation states with the regional flat area payments defined as the national payment ceiling divided by the total eligible area. All other agricultural policies under the Regional Flat Rate scenario are the same as those applying under the baseline.

5.2.3 EU-Wide Flat Rate Scenario

Under the current CAP there are large differences between the average payment entitlements per eligible hectare paid to farmers in different Member States. The alleged inequity of the current distribution of direct income support payments per hectare across the EU has led to suggestions that an EU wide flat rate payment should be introduced instead of the historic and regional payment models in place currently. These differences and the increasing absence of a credible basis on which to justify them is one of the factors motivating the proposed changes that are analysed in the EU-Wide Flat Rate (EFR247) scenario. Under the EFR247 scenario the impact of the introduction of an EU-wide flat area payment set at the level of the EU average per hectare entitlement on full implementation of the CAP in the new Member States is analysed. The per hectare entitlement (of €247 per hectare) is calculated as the sum of all Member States' national ceilings divided by the sum of all Member States' eligible areas. All other agricultural policies under the EU-wide Flat Rate scenario are the same as those applying under the baseline scenario.

5.2.4 *Reduced EU-Wide Flat Rate Scenario*

There are several proposals in the CAP reform literature (e.g. Bureau and Mahé 2008) that argue that the level of EU-wide area payments should be reduced significantly from their existing level. Under the reduced EU-wide flat rate (EFR100) scenario the rate of direct income support payment is set at €100 per eligible hectare and the eligible area is set equal to the agricultural area on which entitlements (under the SPS and SAPS) were established. Under this scenario the rate of modulation is set to zero. The Pillar I funds that are released through the reduction in the level of the average direct income support payment per hectare are used to fund rural development (Pillar II) and other non-agricultural EU policies. All other elements of agricultural policy would remain the same as in the Regional Flat Rate scenario.

5.2.5 *SPS Abolishment Scenario*

The final alternative EU agricultural policy scenario analysed using the AGMEMOD model is the most extreme. It examines what the impact on EU agricultural production and commodity markets would be of a gradual abolition of all direct income supports (both SPS and SAPS). Under the scenario the regional flat rate payments established in the RFR scenario would be reduced to zero in a linear fashion over a 10 years period beginning in 2010 (ZeroSPS scenario).

The RFR and EFR247 scenarios analyse the impact of changing the EU direct income support schemes within a fixed CAP budgetary framework. The EFR100 and ZeroSPS scenarios analyse the impact on agricultural commodity markets of radical changes to the budgetary framework within which the CAP Pillar I measures operate. The scenarios analysed can be ranked on the basis of the departure they represent from the status quo ante that is represented by the baseline scenario, with the RFR scenario being the least radical policy change and the ZeroSPS scenario representing the most radical policy assessed. Table 5.1 summarizes the four scenarios analysed.

5.3 Scenario Results

5.3.1 *Grains and Oilseeds*

On average in EU Member States direct supports under the baseline are projected to account for 19% and 16% of the gross returns (euro per hectare) received for producing grains and oilseeds in 2020. Table 5.2 summarizes the impact of the four scenarios analysed on the gross returns of grains and oilseeds in comparison with the baseline in the EU27 and two sub-EU regions comprising of so-called old Member States and new Member States that acceded to the EU in 2004.

Table 5.1 Scenarios definitions

	SP scheme	Description	Calculation rule
RFR	Regional flat rate per hectare	Move towards regional flat rate entitlements applied to all eligible area. Full decoupling. Modulation as in baseline.	Total level of reference payments of the historical scheme in a region divided by total eligible area in this region.
EFR247	EU-wide flat rate per eligible hectare	The same flat rate (€247/ha) payment entitlement per eligible hectare applied to all EU states. Full decoupling. Modulation as in baseline.	Total annual financial envelope of EU27 Member States divided by total eligible agricultural area of EU27 Member States.
EFR100	EU-wide flat rate per eligible hectare	The same flat rate (€100/ha) payment entitlement per eligible hectare applied to all EU states. Full decoupling. No modulation from 2014.	Fixed value of €100 /ha, with Pillar I budget equal to the product of the fixed per hectare payment times the total EU27 eligible area.
ZeroSPS	SPS abolishment	Gradual reducing budget national ceilings for single farm payments. No modulation from 2014.	Abolition of the single farm payments by 2020 through a series of linear reductions beginning in 2014.

Table 5.2 Impacts on crops gross returns compared to baseline in EU, 2020

	RFR (%)	EFR247 (%)	EFR100 (%)	ZeroSPS (%)
Grains				
EU15	−2.9	−4.7	−7.0	−12.1
EU10	1.0	2.2	−10.0	−19.1
EU27	−1.9	−2.9	−7.4	−13.0
Oilseeds				
EU15	−3.2	−6.4	−7.7	−14.8
EU10	0.1	2.8	−0.9	−3.9
EU27	−2.3	−4.2	−6.1	−12.3

Source: AGMEMOD 4.0 (2010)

The Regional flat rate (RFR) and EU-wide flat rate scenarios (EFR247) are projected to lead to relatively small changes in the amount of income support per hectare versus the baseline scenario. Given the small magnitude of the projected changes in gross returns, the changes in total cereal area harvested in the EU27 are expected also to be small. The decline in the total cereals area harvested in 2020 relative to the baseline is larger under the EFR247 scenario than under the RFR scenario. Under the EFR247 there is a redistribution of budgetary envelopes between the Member States and larger changes in support per hectare than under the RFR scenario. As a result total cereal area declines by 1.2% relative to the baseline.

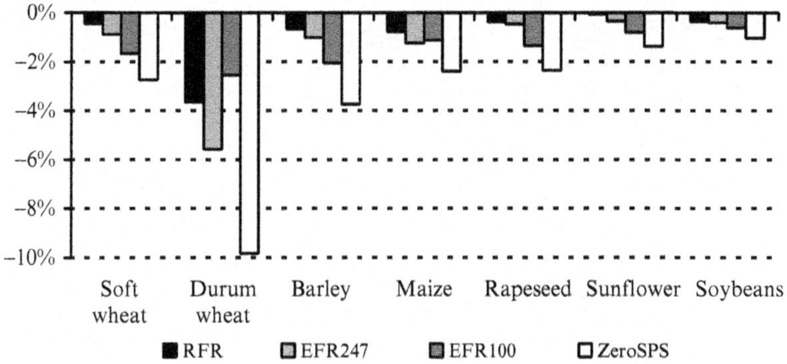

Fig. 5.1 Impacts on EU crops areas compared to baseline, 2020 (Source: AGMEMOD 4.0 2010)

The decline in cereals area harvested, when all Member States apply a regional flat area payment system, is projected to be smaller than under the EFR247 scenario, with the area projected to be 0.8% lower in 2020 than under the baseline.

Under the more radical EFR100 and ZeroSPS scenarios, in which the total EU budget for direct payments is reduced, the change in total EU27 cereal area harvested (relative to the baseline) is larger than under the RFR and EFR247 scenarios. Nevertheless the magnitude of the projected changes in EU27 cereal area harvested is still relatively small. Under the EFR100 scenario, the total cereal area is projected to decline by 1.9% relative to the baseline. The total cereals area harvested projected under the ZeroSPS scenario, where the single farm payment is abolished over the period to 2020, is 3.6% lower than under the baseline by 2020. The magnitude of the changes in individual commodity culture areas harvested is projected to differ somewhat from the projected change in total cereal area, and Fig. 5.1 shows the scenario impacts for individual cereal and oilseed areas.

In general the declines in oilseed area harvested under all four of the reform scenarios analysed are smaller than the changes projected to occur in cereal areas harvested under these scenarios. The ranking of the four scenarios in terms of impact on oilseed area harvested is the same as the ranking of the scenarios in terms of their projected impact on cereal markets, the largest change is projected to occur under the ZeroSPS scenario, where total oilseed areas harvested is projected to decline by 2.8% relative to the baseline. The smaller impact of the scenarios analysed on oilseed area harvested is due to the smaller share of total per hectare margins that are accounted for by direct payments when compared with cereal crops.

There are differences in the impact of the considered scenarios on the supply of different individual cereal and oilseed crops (see Fig. 5.2). The impact on all the crops modelled is projected to be negative and significantly more negative under the ZeroSPS scenario than the other reform scenarios analysed. Durum wheat among the cereals modelled and rapeseed among the oilseed crops modelled are exceptional in respect of the magnitude of the negative impact of the scenarios analysed.

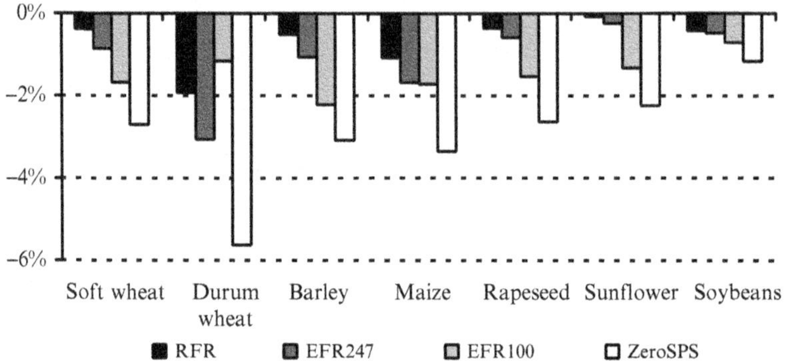

Fig. 5.2 Impacts on EU crops production compared to baseline, 2020 (Source: AGMEMOD 4.0 2010)

The large decrease in durum wheat and rapeseed harvested areas is due to the larger proportion of the gross return per hectare from rapeseed and durum wheat production that is projected to be accounted for by subsidies under the baseline.

The large change in durum wheat production also reflects the fact that Member States that produce durum wheat, on average, receive direct payments per hectare that are significantly greater than the EU average per hectare payment. Payments per hectare to durum wheat producers are also, on average, larger than payments made to other farmers in those Member States. The move to flat area payment models (such as under the RFR and EFR247 scenarios) significantly reduces the incentive price for durum wheat in those Member States where durum wheat is produced. Member States where the incentive prices for grains increase under the reform scenarios analysed (e.g. Latvia) in general are agronomically unsuited to durum wheat production.

The four scenarios analysed are projected to lead to a reduction in the production of grains relative to the baseline (see Fig. 5.2). The reductions in the production of soft wheat, durum wheat, maize and barley, largest under the SPS abolishment scenario, are still relatively minor. In contrast to the cases of the other grains modelled, under all scenarios the reduction in durum wheat harvested area is offset by projected increases in the yield per hectare. The impact of the four scenarios analysed on the EU27 oilseed production are also projected to be limited. The production of all oilseeds under the most extreme policy reform analysed only contracts by 3% relative to the baseline level(see also Box 5.1).

The limited impact of the scenarios analysed on the EU27 supply of cereals and oilseeds means that the impact of the scenarios analysed on grain and oilseed (and meal and oil) prices is also very limited. The projected contraction in grain and oilseed supply leads to very small increases in prices on all markets with the exception of the soybean market, where prices are projected to remain unchanged under all four of the alternative scenarios (Fig. 5.3). Grain and oilseed prices are projected to increase on average by less than 1% under all policy scenarios considered.

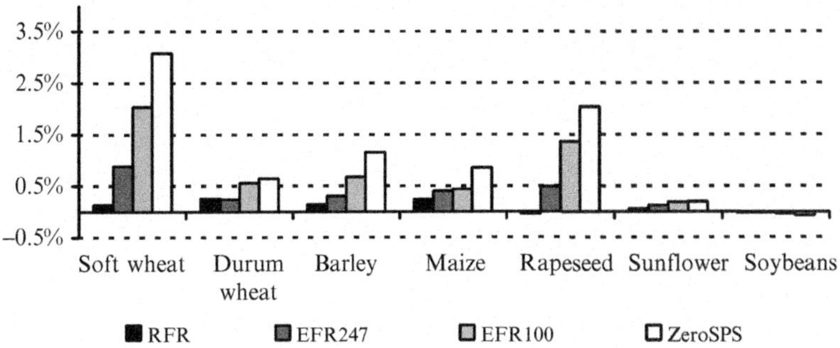

Fig. 5.3 Impacts on EU crop prices compared to baseline, 2020 (Source: AGMEMOD 4.0 2010)

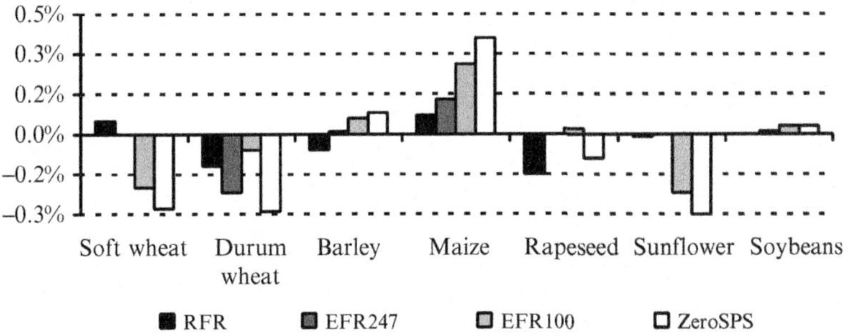

Fig. 5.4 Impacts on EU crops domestic uses compared to baseline, 2020 (Source: AGMEMOD 4.0 2010)

The impacts of the four scenarios analysed on the domestic use of grain and oilseeds (Fig. 5.4) are very minor with changes of less than a 0.1% relative to the baseline projected under all scenarios analysed.

5.3.2 *Livestock and Meat*

On average in EU Member States direct supports under the baseline are projected to account for 8%, 0% and 25% of the reaction price (euro per 100 kg) for beef and veal, broilers and sheep meat respectively by 2020. Table 5.3 summarizes the impact of the four scenarios analysed on the reaction prices of these three meat categories when compared with the baseline for the EU27 and two sub-EU regions comprising of so-called old Member States and new Member States that acceded to the EU in 2004.

The impact on livestock markets differ across the four reform scenarios analysed and the scenarios have much more diverse impact on the four livestock and meat markets modelled than was the case with grains and oilseeds. This greater impact reflects the greater dependence of some of these sectors on coupled direct payments

Box 5.1 Selected Member States, Scenario Impacts for Grains and Oilseeds

The impact of the alternative policy scenarios on grains and oilseeds markets varies between different EU Member States. The following graphs show the projected impact on production of soft wheat and rapeseed in France and the Netherlands, where historical payment rights prevail, and in Germany and Poland, where flat area payment systems are used.

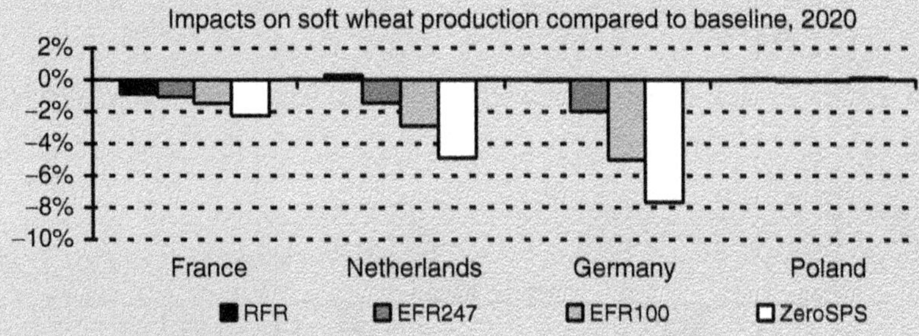

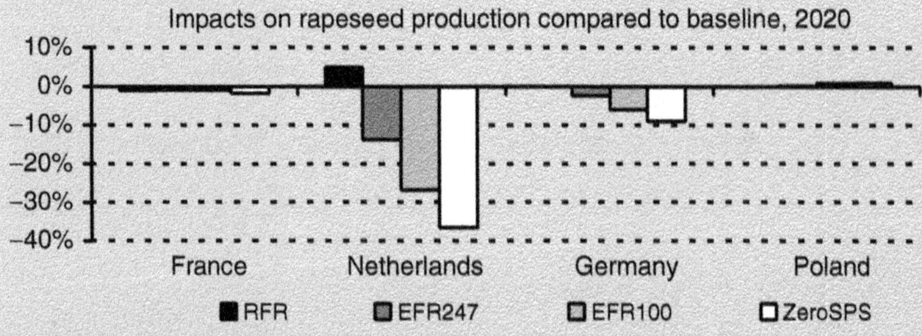

The ranking of the impacts of the four policy scenarios at the individual EU Member State level is the same as that at the EU aggregate level, though there are notable differences in the magnitude of the projected impacts of the reforms between Member States. In Poland the level of payments per hectare under the baseline is relatively low. This means that the impact of the changes to direct payment systems analysed on Polish soft wheat and rapeseed production is negligible. The impacts, under all scenarios, on French, Dutch and German soft wheat markets are as expected, with German production of soft wheat projected to decline by the greatest percentage relative to the baseline.

The impacts of the alternative policy scenarios analysed on rapeseed production are somewhat more diverse. The size of German production effects is similar to that projected for soft wheat. The very strong projected impact in the Netherlands is due a scale effect. With only limited rapeseed production in the Netherlands under the baseline even minimal absolute changes are reflected in large percentage changes.

Table 5.3 Impacts on livestock price returns compared to baseline in EU, 2020

	RFR (%)	EFR247 (%)	EFR100 (%)	ZeroSPS (%)
Beef and veal				
EU15	−2.8	−3.0	−3.2	−6.0
EU10	−1.1	0.9	−4.0	−7.9
EU27	−2.0	−1.3	−3.5	−6.5
Broiler				
EU15	0.4	0.6	0.7	0.9
EU10	0.5	0.6	0.8	1.1
EU27	0.5	0.6	0.7	1.0
Sheep meat				
EU15	−10.0	−10.4	−11.1	−17.9
EU10	−5.6	−0.3	−16.8	−29.6
EU27	−8.0	−6.1	−13.0	−21.1

Source: AGMEMOD 4.0 (2010)

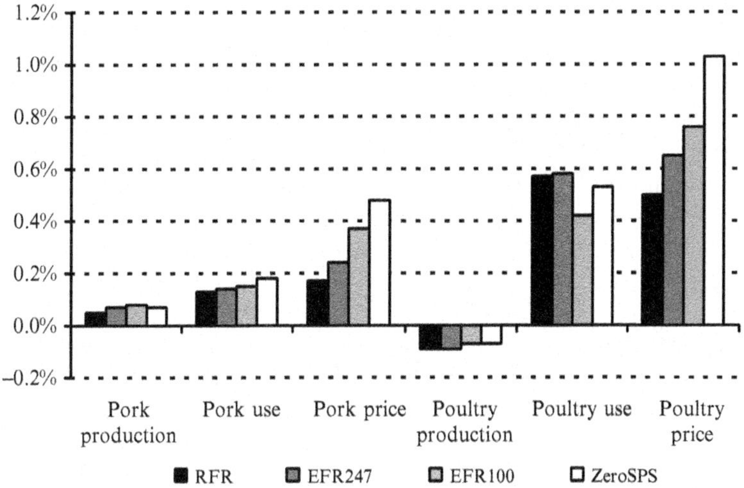

Fig. 5.5 Impacts on EU pig and poultry sectors compared to baseline, 2020 (Source: AGMEMOD 4.0 2010)

and the concentration of production of these commodities in Member States with larger than average direct income support per hectare.

The impacts of the four scenarios on the pig meat and poultry meat sectors are relatively minor. Farmers in these sectors are generally not in receipt of direct income support payments and the impact of the scenarios analysed arise largely from cross commodity market impacts on input (feed) costs and on the competitiveness of these (white) meats versus red meats (beef and sheep meat). The greater negative impact of the four scenarios analysed on EU27 cattle and sheep production reflects the greater dependence of these activities on direct payments and income supports from the CAP Pillar I. Thus, the exposition of results that follows focuses on the scenario results for the cattle and sheep sectors (Fig. 5.5).

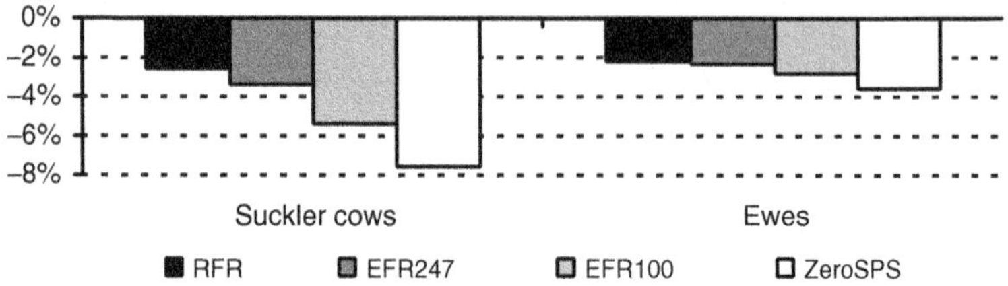

Fig. 5.6 Impacts on EU suckler cows and ewes numbers compared to baseline, 2020 (Source: AGMEMOD 4.0 2010)

Under the regional flat rate (RFR) and EU-wide flat rate (EFR247) scenarios in which the current budgetary framework is maintained, remaining coupled payments are fully decoupled. Under both of these scenarios, EU27 suckler cow numbers decline relative to the baseline. By 2020, under the RFR scenario, EU suckler cow numbers are projected to be 2.4% lower than under the baseline. Under the EFR247 scenario the reduction in the suckler cow herd is larger, with numbers projected to be 3.4% lower than under the baseline. Most of the EU suckler herd is farmed in the EU15 Member States and on average these Member States have higher per hectare single farm payments than the EU12 Member States. The impact of the decoupling of payments that under the baseline had remained coupled to production is also significant. Suckler cow numbers are projected to increase in some new Member States, though it should be noted that the projected increases are from a very low base and thus the associated changes in absolute cow numbers are very small (Fig. 5.6).

Under the EFR100 and ZeroSPS scenarios, with reduced budgetary resources, the suckler cow numbers in the EU27 are projected to decline. The magnitude of the projected decline is greater under these two scenarios than under the RFR and EFR247 scenarios. Under the EFR100 scenario, EU27 ending stocks of cows decline by over 5%, while under the more extreme ZeroSPS scenario the ending stocks of suckler cows in the EU27 in 2020 decline by 7% relative to the baseline. As under the previous RFR and EFR247 scenarios, the decline in suckler cow numbers in some Member States is greater than that projected for the overall EU27.

Figure 5.7 graphs the impacts of the four scenarios analysed on beef and sheep meat markets. The projected impact of the scenarios analysed on the beef markets differs by Member State, depending on whether or not the beef production in that Member State is primarily based on offspring from the dairy cow herd or on a mix of the off-spring of suckler cow and dairy cow herds. In the EU27, approximately two thirds of the cows are dairy cows, so that the majority of beef production is based at least in part on dairy farmers' decisions, which will be influenced primarily by dairy policy and expected development in real milk prices. In some Member States such as Ireland, France and the UK the suckler cow herd is of more importance, though in no EU Member State do suckler cow numbers exceed dairy cow numbers.

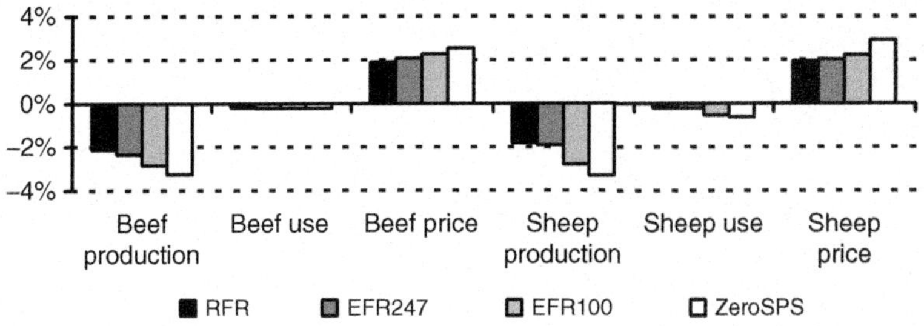

Fig. 5.7 Impacts on EU beef and sheep sectors compared to baseline, 2020 (Source: AGMEMOD 4.0 2010)

Under all scenarios analysed, total EU27 production of beef decreases as a result of the projected contraction of the EU suckler cow herd, the magnitude of the contraction in beef production is lower than the change in suckler cow numbers due to the role of the dairy sector in beef production. The total domestic use of beef in the EU27 under all scenarios is projected to decline marginally as a result of the increase in beef prices that are projected to arise as a result of the projected contraction in beef production from the suckler cow herd. The EU27 beef net import position under all of the scenarios increases relative to the baseline.

The EU27 sheep meat sector is also negatively affected under all of the policy reform scenarios analysed. Under the RFR and EFR247 scenarios, all direct payments are fully decoupled and this change and the flattening of direct payment models is projected to lead to a decline in the EU27 ewe flock and volume of lamb production relative to the baseline. By 2020, under the RFR scenario, EU27 ewe numbers are projected to decline by 3.1% relative to the baseline, while under the EFR247 scenario ewe numbers are projected to be marginally lower again. The projected small scale of the negative adjustment in EU27 ewe numbers is partly due to the importance of the Southern European light lamb-milk sheep production system which is somewhat less dependent on coupled supports than the heavy lamb production systems of North Western Europe. The impact on the EU27 sheep sector of the EFR100 and ZeroSPS scenarios, where the budgetary resources devoted to the CAP Pillar I are reduced, are an amplification of the negative impacts of the two first scenarios analysed. The reduction of budgetary support to agriculture leads to a contraction in the EU27 ewe breeding stock relative to the baseline. By 2020, under the EFR100 scenario the number of ewes in the EU27 is projected to 3.7% lower than under the baseline, while it is projected to be 4.6% lower under ZeroSPS scenario.

This decline in sheep breeding numbers leads to a concomitant decline in the volume of lamb meat produced in the EU when compared with the EU27 baseline projections. Lamb production contracts under each of the alternate policy scenarios. Lower domestic use of lamb in the EU27 is projected as a result of the increased price of lamb that is expected to arise. Overall the EU imports of lamb are projected to grow relative to the baseline (see also Box 5.2).

Box 5.2 Selected Member States, Scenario Impacts for Beef and Sheep

The impact of alternate policy scenarios on beef and sheep production varies between EU members. The graphs below show the results for beef and sheep production in France and Ireland, where historical direct payments scheme prevails, and in Slovenia which applies a hybrid SPS scheme and in Latvia where the SAPS has been applied.

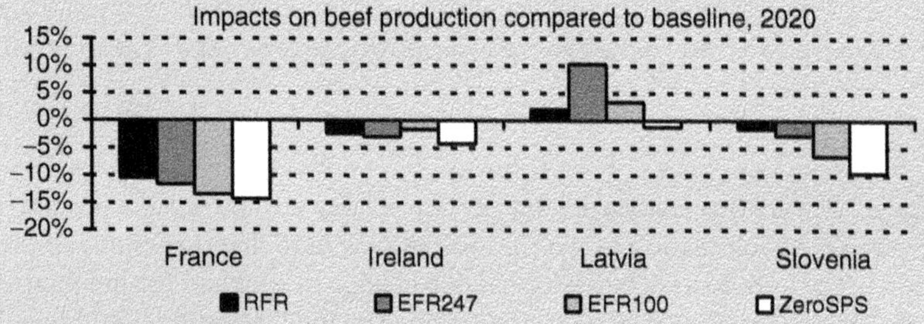

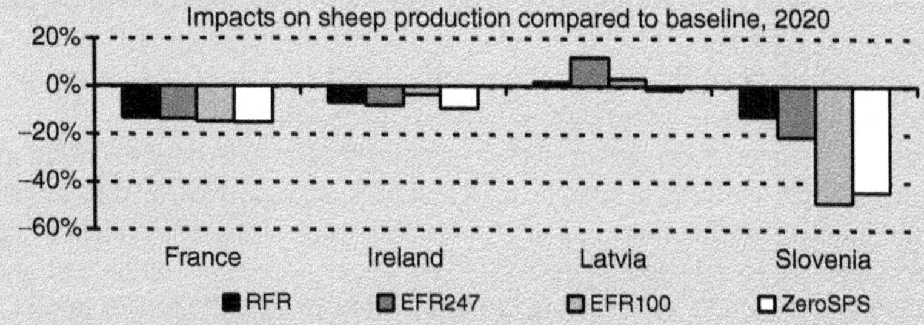

The Member State level impacts of the four scenarios are very different. The differences across the EU Member States result from the differences in the implemented payment schemes and the larger differences in average payment per hectare under the baseline. The projected impact of the reforms on those Member States applying historical and hybrid direct payments schemes (France, Ireland and Slovenia) is stronger than the impact of the reforms on the Member State that uses the SAPS scheme (Latvia). Under the baseline suckler cow premiums remain fully coupled in France and ewe premiums remain partly coupled. Under all four policy scenarios, payments are fully decoupled from production and this is projected to result in a significant decrease in French beef and sheep production.

Under the EFR247 scenario, which redistributes currently agreed budgetary resources amongst the member states, the direct payment level is projected to increase for countries such as Latvia. This is projected to lead to higher beef and sheep meat production.

Table 5.4 Impacts on milk price returns compared to baseline in EU, 2020

	RFR (%)	EFR247 (%)	EFR100 (%)	ZeroSPS (%)
EU15	–0.5	–0.6	–0.6	–1.6
EU10	0.2	1.0	–0.9	–2.0
EU27	–0.2	0.0	–0.8	–1.8

Source: AGMEMOD 4.0 (2010)

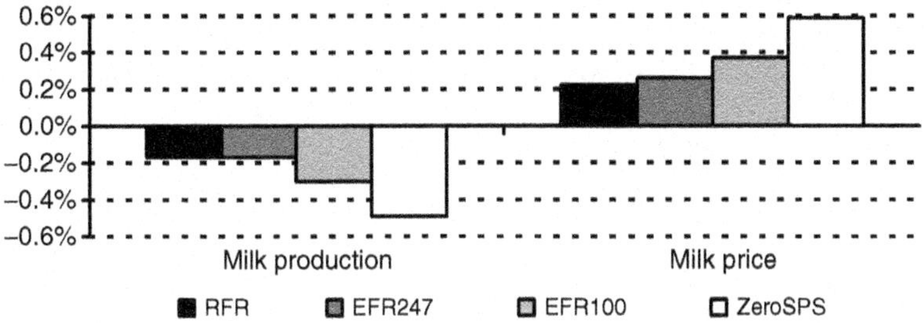

Fig. 5.8 Impacts on EU milk market compared to baseline, 2020 (Source: AGMEMOD 4.0 2010)

5.3.3 *Milk and Dairy Products*

On average in EU Member States direct supports under the baseline are projected to account for 3% of the price returns (euro per 100 kg) for milk in 2020. Table 5.4 summarizes the scenario effects on the price returns of milk compared to the baseline for the EU27 and two sub-EU regions comprising of so-called old Member States and new Member States that acceded to the EU in 2004.

The increase in milk quota is taken up in a number of EU Member States under the baseline scenario, but it is found that due to successive milk quota increases, the milk quotas are no longer binding in many Member States in advance of their expiration in 2015. The impacts of the alternate policy scenarios analysed on EU milk and dairy markets are very small, with projected reductions in milk production of the order of 0.15–0.4% at EU27 level depending on the scenario analysed. It is notable that these projected changes in production under the reform scenarios are the sum of both positive and negative Member States changes in production relative to the baseline, i.e. growth in production in some Member States offsets contraction in others.

Figure 5.8 summarizes the EU outcome for the milk price and milk production under the four alternate policy scenarios analysed relative to the baseline. By 2020 the EU milk price increase is projected to be between 0.3% and 0.7% higher depending on the scenario. The production of milk is projected to contract at the EU27 aggregate level due to the reduction in the level of reaction prices as a result of the changes to the direct payment regimes, but the magnitude of the projected change relative to the baseline is very small.

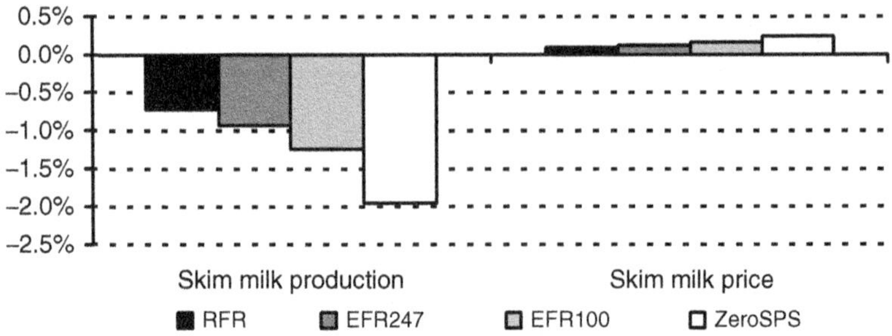

Fig. 5.9 Impacts on EU skim milk powder market compared to baseline, 2020 (Source: AGMEMOD 4.0 2010)

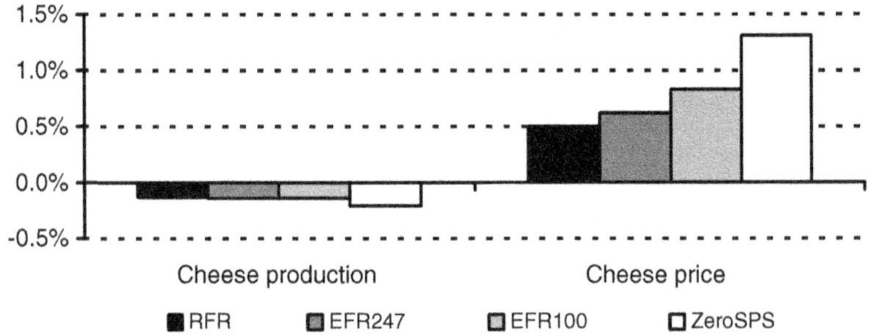

Fig. 5.10 Impacts on EU cheese market compared to baseline, 2020 (Source: AGMEMOD 4.0 2010)

The impacts of the alternate policy scenarios on skim milk powder and cheese are given in Figs. 5.9 and 5.10. The move to flat area payment systems (RFR and EFR247 scenarios), particularly in EU15 Member States, reduces the policy support relative to the baseline. The introduction of a €100/ha EU wide flat area payment (EFR100) and the abolishment of the single farm payment under the ZeroSPS scenario further reduces the level of policy support in most Member State models.

Due to the reduced volume of milk production under each of the four reform scenarios, the prices of dairy products are higher than under the baseline. The largest increase in prices is that projected to occur on cheese markets, where by 2020 the price of cheese is expected to be 1.2% higher under the ZeroSPS scenario than under the baseline.

Under each of the alternate policy scenarios analysed, the production of all dairy products decreases relative to the baseline, with the largest percentage decrease projected to occur in the production of milk powders. In general, however, the simulated changes in production, use and prices of dairy commodities are very minor. The minor scale of the impact of the reform scenarios on milk and dairy markets is due to the low share of direct payments in the revenues of the majority of dairy farms (see also Box 5.3).

Box 5.3 Selected Member States, Scenario Impacts for Milk

The impact of alternate scenarios policy on the dairy market varies amongst EU members. This box focuses on the scenarios impacts on the production of milk in various Member States. The graph shows the results for the milk production in France and the Netherlands, where historical direct payments schemes prevail under the baseline, in Germany with its regional payment scheme and in Poland where the SAPS has been applied.

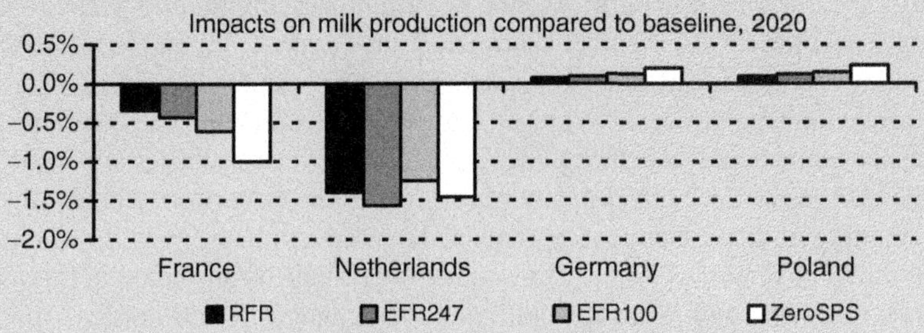

The results show moderate scenarios impacts. Negative impacts on the French and Dutch milk production are projected. This development is caused by the reduction in policy support in these countries under the alternative policy scenarios analysed.

5.4 Conclusion

The discussions and negotiations concerning the future of the CAP direct payments system, wider reform of the CAP pillar I and II measures and parallel negotiations on the determination of new EU multi-annual financial framework for the period 2014–2020 started in 2010. These negotiations will mean that analysis of the impact of agricultural policy changes at Member State and EU levels will increasingly be in demand. Changes to the level of the budgetary support to EU agriculture, changes in the method of distributing such budgetary support between farmers and amongst Member States, and changes in the distribution of support between Pillar I and Pillar II measures are all possible and probable over the medium term.

The scenarios examined using the AGMEMOD model that have been presented in this chapter involved the full decoupling of all remaining coupled direct payments and (1) a movement to a regional payment model in all Member states and (2) a movement to an EU wide flat area payment model under three different budgetary options. In the first EU wide flat area payment scenario the budget devoted to agricultural income support (Pillar I of the CAP) was maintained but redistributed amongst EU Member States. In the second flat area payment scenario the value of payments were over a transition period, reduced (or in some cases increased) to €100 per eligible hectare, while in the third scenario direct income supports were phased out over the upcoming 7 year multi-annual financial framework.

The results presented in this chapter illustrate the complexity of the implementation of existing EU agricultural policy across different Member States and illustrate the usefulness of the AGMEMOD model, with its Member State detail, in analysing changes in the EU agricultural policy environment.

The complexity of the reform scenarios analysed in the European context arises from (i) the differing nature of baseline or current agricultural policy environments that exist in different EU member states under the current CAP, and (ii) from the fact that policy changes and associated market changes in one member state can affect the market environment projected to prevail in other Member States. The contrasting supply responses under the alternate policy scenarios between countries are apparent.

The alternative CAP direct payment system reform scenarios examined involved the flattening of direct payment systems and the reduction and even the elimination of the direct payments systems. The results of the analysis presented in this chapter indicate that such significant agricultural policy changes would not lead to radical changes in EU27 agricultural production at the aggregate level. The projected impact of both the Regional Flat Rate scenario and the EU wide Flat Rate scenario underline the importance of the baseline or *status quo ante* policy position of different Member States in assessing the impact of a common policy change. The project impact of the alternative policy reform scenarios at the individual Member State level, suggests that the impact of the considered reforms is greater than at the level of the aggregated EU. The products that are projected to be most affected by the reforms analysed are the beef and sheep meat where coupled direct payments continue to affect the decision making process of farmers in many Member States and where the changes in the value of subsidies per sector are the most significant. The projected changes in grain and oilseeds production are only significant in the case of a drastic reduction of direct payments support (the ZeroSPS scenario).

The introduction of an EU-wide flat area payment is projected to lead to increases in production in those Member States where the level of direct income support in 2013 is lower than the average EU level. In contrast, levels of agricultural activity and associated agricultural commodity production are projected to decline in those Member States with above average direct payment receipts per hectare. However, as noted above, under the EU flat rate scenarios analysed there are no dramatic changes projected in the pattern of EU agricultural production. These results suggest that a bolder and perhaps more horizontally equitable decoupled payment regime could be considered in the current CAP reform process.

The two budgetary related scenarios (EFR100 and ZeroSPS) examined the commodity market impacts of policy changes that would significantly reduce the budgetary resources devoted to CAP Pillar I measures. While the commodity market impacts of these two scenarios were the largest of the four CAP policy alternatives analysed, the magnitude of the impacts on EU agricultural commodity markets and on agricultural output prices were relatively modest. The greatest impacts, as anticipated by Bureau and Mahé (2008), were on the specialized beef and sheep sectors. In the Member States which, under the baseline, already have a regional flat area payment system (the majority of the new Member States, Germany and some other old Member States) beef production is projected to be only marginally affected by a move to a regional payment model. The projected decline in the EU production is as a result of the negative impact of the regional flat area payment model on beef production in Member States such as France, Italy and the Netherlands. Under the baseline, France retains coupled beef premiums and utilizes a historical decoupled payments model. As a result, the production of beef is projected to decline under the regional flat area scenario.

The large percentage increase in beef production in some new Member States such as Poland, the Baltic States and Bulgaria, under the EU wide flat area payment scenario contrasts with the minor impact projected under the regional flat rate scenario. Budgetary support per hectare in the majority of the new Member States under the baseline is significantly lower than in the EU15 and the move to an EU wide flat rate dramatically increases the support provided and leads to an increase in beef production in these parts of the EU. In contrast, the move to an EU wide flat rate payment model under the EU wide flat area payments scenario significantly reduces the budgetary support to Member States with the historical model of direct payments like France, Ireland and Italy.

Finally, this chapter illustrates the importance of modelling all markets simultaneously within a framework, such as that provided for in the AGMEMOD model, that has the flexibility of incorporating additional Member States and/or commodity markets. The results presented illustrate the usefulness of the AGMEMOD model as a tool with which to examine alternative EU agricultural policy proposals. The discussion in this chapter has largely focused on the analytical results at the level of the EU27, however, one of the advantages of the AGMEMOD tool is its ability to illustrate the differing nature of the impacts of policy changes across the European Union's increasingly heterogeneous set of Member States. The examples presented in this chapter's text boxes provide an illustration of the AGMEMOD model's rich capacity to examine the differential Member State impact of changes in one of the European Union's most important *common* policies.

References

Begg I (2005) Funding the European Union. A federal trust report on the European Union's budget. The Federal Trust for Education and Research, London

Begg I, Heinemann F (2006) New budget, old dilemmas. Briefing Note, Centre for European Reform, London

Begg I, Sapir A, Eriksson J (2008) The purse of the European Union: setting priorities for the future. Swedish Institute for European Policy Studies, Stockholm

Buckwell A (2007) Next step in the CAP reform. EuroChoices 6(2):13–19

Bureau JC, Mahé LP (2008) CAP reform beyond 2013: an idea for a longer view. Notre Europe Studies and Research No. 64. Notre Europe, Paris

Burfisher ME, Hopkins J (2003) Decoupled payments: household income transfers in contemporary U.S. agriculture. Agricultural Economic Report No. 822, United States Department of Agriculture (USDA), Economic Research Service (ERS), Washington, DC

Chau NH, de Gorter H (2005) Disentangling the consequences of direct payment schemes in agriculture on fixed costs, exit decisions, and output. Am J Agr Econ 87(5):1174–1181

Cipriani G (2007) Rethinking the EU budget: three unavoidable reforms. Centre for European Policy Studies (CEPS), Brussels

Daugbjerg C, Swinbank A (2007) The politics of CAP reform: trade negotiations, institutional settings and blame avoidance. J Common Market Stud 45:1–22

ECORYS (2008) A study on EU spending. Final Report. Commissioned by the European Commission, Directorate General for Budget, Contract No. 30-CE-0121821/00-57. ECORYS Nederland BV, Rotterdam

Esposti R (2008) Reforming the CAP: an agenda for regional growth. 109th seminar of the European association of agricultural economists. Viterbo, Italy, 20–21 Nov 2008

Garzon I (2006) Reforming the common agricultural policy: history of a paradigm change. Palgrave Macmillan, Houndmills

Grybauskaitė D (2008). Reforming the budget, changing Europe: results of the public consultation. Plenary presentation at budget conference "Reforming the Budget, Changing Europe", Brussels, 12 November 2008. Available via http://ec.europa.eu/budget/reform/library/conference/intro_grybauskaite.pdf. Cited 21 June 2011

Hennessy DA (1998) The production effects of agricultural income support policies under uncertainty. Am J Agr Econ 80:46–57

OECD (2001) Decoupling: a conceptual overview. Organisation for Economic Cooperation and Development (OECD), Paris

OECD (2003) Multifunctionality – towards an analytical framework. Organisation for Economic Cooperation and Development (OECD), Paris, Paris

Roe T, Somwaru A, Diao X (2003) Do direct payments have intertemporal effects on US agriculture? In: Moss CB, Schmitz A (eds) Government policy and farmland markets. Iowa State University Press, Ames

Salputra G, Chantreuil F, Hanrahan KF, Donnellan T, Van Leeuwen M, Erjavec E (2011) Policy harmonized approach for the EU agricultural sector modelling. Agric Food Sci 20(2): 119-130.

Swinnen JFM (2008) The perfect storm – the political economy of the Fischler reforms of the common agricultural policy. Centre for European Policy Studies (CEPS), Brussels

Yrjölä T, Kola J (2004) Consumer preferences regarding multifunctional agriculture. Int Food Agribusiness Man Rev 7(1):78–90

Chapter 6
Conclusions and Future Work

Frédéric Chantreuil, Kevin F. Hanrahan, and Myrna van Leeuwen

In developing the AGMEMOD model of European agricultural commodity markets the AGMEMOD Partnership has shown that it is possible to establish an econometric, dynamic, multi-product partial equilibrium model of European agriculture based on a "bottom up" approach. A total of 27 country level dynamic, multi-product partial equilibrium models of agriculture have been developed and integrated to form a coherent partial equilibrium model of EU27 agriculture. This model is used to generate (baseline) agricultural commodity market outlook information and analysis and to conduct policy change impact (scenario) analysis.

The country sub-models within the AGMEMOD model are a product of the research conducted by teams in the respective countries. In addition country level market experts have contributed to the development and validation of these models. This modular structure allows the AGMEMOD model to reflect the inherent heterogeneity of both EU Member State agriculture and EU CAP implementation. This country level distinctiveness, an intrinsic characteristic of the AGMEMOD model, is necessarily balanced by the requirement to integrate the models in a pan-EU model setting. Model integration has been ensured through the adherence by the country model development teams to the modelling templates and validation procedures that were outlined in Chap. 2. The AGMEMOD baseline and scenario analysis presented

F. Chantreuil (✉)
Université de Caen Basse Normandie, UFR Sciences Economiques et Gestion,
17 rue Claude Bloch, BP 5186, F-14032 Caen cedex, France
e-mail: Frederic.Chantreuil@unicaen.fr

K.F. Hanrahan (✉)
Teagasc, Rural Economy Development Programme, Agricultural Economics
and Farm Survey Department, Athenry, Galway, Ireland
e-mail: Kevin.Hanrahan@teagasc.ie

M. van Leeuwen (✉)
Research Area International Policy, LEI Wageningen UR, Alexanderveld 5,
2585 DB The Hague, The Netherlands
e-mail: Myrna.vanleeuwen@wur.nl

F. Chantreuil et al. (eds.), *The Future of EU Agricultural Markets by AGMEMOD*,
DOI 10.1007/978-94-007-2291-0_6, © Springer Science+Business Media B.V. 2012

in Chaps. 4 and 5, along with the detailed description of its structure, illustrate the capacity of the model to undertake meaningful and insightful analysis.

Aside from the development of an operational model of EU agriculture, the AGMEMOD project has also contributed to the establishment of a rich set of agricultural economics research networks within individual Member States and across the expanded EU. At Member State level, partners have liaised with industry experts in the evaluation of their country model's performance. This evaluation process has also served as a form of dissemination activity and has contributed to awareness of the research undertaken with the support of Community funds. Moreover, expert evaluation reinforces the credibility of the analytical use of the model in policy analysis and support.

The project has also contributed to the establishment of a strong and vibrant pan-European network of research economists working in the field of agricultural policy analysis. A network with this capacity would have been difficult to establish through any another means. The success of the network is reflected in the numerous conference papers and jointly authored, peer reviewed articles produced.

The number of researchers participating in AGMEMOD has increased over time, increasing the communication and coordination challenge. The development of the policy harmonisation approach described in Chap. 2 and the creation of the technical modelling tools presented in Chap. 3 were vitally important in addressing these challenges. These important innovations enhanced the model's transparency and consistency and facilitated the expansion of its analytical scope and have allowed new commodities and countries to be added in a practical fashion. Using the AGMEMOD standardised modelling structure, EU candidate or potential candidates to accession (Croatia, Turkey and Macedonia) and EU neighbours (Russia and Ukraine) have been modelled and are part of the current AGMEMOD 4.0 model. This expansion of the geographical coverage allows us to provide baseline outlooks for these non-EU countries and offers the possibility to examine questions relating to further EU expansion or bilateral trade scenarios. Further expansion of the model's geographic coverage is ongoing with the incorporation of models of agricultural commodity markets in countries such as Kazakhstan, China and Brazil currently underway.

In the medium term, several other important areas of AGMEMOD model development can be identified. One will involve the endogenisation of world agricultural commodity price formation within the AGMEMOD model framework. Currently world agricultural commodity market prices are exogenous to the AGMEMOD model. Projections of baseline and alternative policy scenario agricultural commodity prices are obtained from other partial equilibrium models. With the endogenisation of world price formation within the AGMEMOD modelling framework the model will be capable of addressing the impact of policy changes in the EU on world agricultural commodity market prices and agricultural commodity supply and use outside of the EU.

A second area of further research for AGMEMOD will involve the further development of the modelling of emerging bio-energy markets inside and outside of the EU. On the supply side, agriculture and energy markets have always been related given the important role of energy and energy related products such as chemical fertiliser as inputs into agricultural production. Energy policy initiatives that support the production of bio-energy and the use of bio-energy products through the provision of production and consumption subsides as well as the setting of blending mandates have created new linkages between energy and agricultural markets.

The growth in demand for bio-energy in the EU and internationally may have important implications for agricultural land use, agricultural commodity prices and global food security that may require the more detailed modelling of land and other agricultural factor markets.

Other important and topical areas of model development will be those that address questions concerning the impact of agricultural input and output price volatility on agricultural production and agricultural sector income. This may involve the development of a stochastic element within the AGMEMOD model structure that would allow for the analysis of the impact of shock events on agricultural commodity markets.

The contribution of agricultural production to greenhouse gas (GHG) emissions and the impact of climate change policy and other policies that seek to reduce emissions of GHG on agricultural production will also be important. The AGMEMOD model currently produces projections to a 10 year horizon of the agricultural activity levels that are used to generate sectoral GHG emissions under the auspices of the Intergovernmental Panel on Climate Change (IPCC). To generate Member State level projections of the emissions of GHG from agriculture will require the combination of information on GHG emissions per unit of activity and activity intensity (yields) with an economic model such as AGMEMOD. Such research would allow AGMEMOD to analyse the impact of agricultural policy change on GHG emissions at Member State and EU levels and the impact of climate change policy (at Member State and EU level) on agricultural markets and incomes.

The development of the AGMEMOD model reflects the desire of European policy analysts and policy makers to understand the implications of market and policy developments at both Member State and EU levels. The AGMEMOD model successfully combines the advantages of Member State detail and knowledge within a combined EU model architecture and allows for the integrated analysis of agricultural policy across both commodity and Member State dimensions. The lessons learned in the development of such a model are however likely to be applicable beyond the EU.

Internationally, the emergence and development of regional blocks such as the Southern African Development Community (SADC), Mercado Común del Sur (Mercosur) and the Association of Southeast Asian Nations (ASEAN) means that the experience of the AGMEMOD Partnership may be particularly relevant to agricultural economists in these regions. The EU is a unique experiment in the degree to which economic and political sovereignty has been pooled by its constituent Member States. Within the EU agriculture is particularly unusual in the degree to which Community policy dominates national considerations. While it is unlikely, in the medium term at least, that regional blocks such as SADC and Mercosur will evolve into political entities similar to the EU, it is nevertheless probable that their agricultural and other markets will become increasingly integrated. As barriers to trade with fellow regional block members are lowered analysis at regional and member state level became increasing relevant. The approach adopted by the AGMEMOD Partnership, involving the development of country level analytical capacity, within a link-model architecture, will be of interest to those examining the future of agricultural markets and agricultural developments in such regions.

The support of the EU and the European Commission in particular has been central to the development of both the AGMEMOD network of EU agricultural economists and the AGMEMOD model. The AGMEMOD project has deepened the integration of the European research area in the field of agricultural economics and has supported the development of analytic capacity in each of the Member States. The analysis of agricultural commodity markets, at Member State and EU levels will become increasingly important as European agricultural policy seeks to address the challenges of the twenty-first century. These challenges include assuring global food security and mitigating the impact of economic growth on GHG emissions. Agriculture and agricultural policy will play an important part in addressing these and other issues.

In this book we have argued that the partial equilibrium, bottom up approach to modelling EU agricultural commodity markets has merit. Even as the process of European integration and the application of a common agricultural policy regime seemingly reduce the inherent heterogeneity of European agriculture, many relevant policy questions remain which require a Member State level assessment. The need to achieve political agreement on agricultural policy change between EU Member States, European Parliament and the European Commission will ensure that interest remains in how policy and wider market developments will affect agricultural sectors at the level of the Member State.

The future agenda for agriculture, agricultural markets and policy interactions will increasingly be motivated by what currently appear to be emerging concerns. These concerns include such issues as agriculture's impact on the environment or Europe's contribution to the global food security. Partial equilibrium models such as the AGMEMOD model have the capacity to address such issues in a rigorous and transparent manner this will make such models increasingly important.

Looking to the future, national ministries are likely to increasing require an evidence based decision making framework to support their policy positions. The basis of such support will be the provision of informed baseline and scenario analysis at the Member State and EU levels. The future of the AGMEMOD model will depend on the capacity of the model and those involved in its maintenance and development to address the important future policy questions faced by European agriculture.

Glossary

CAP	Common Agricultural Policy
CGE	Computable General Equilibrium
CNDP	Complementary National Direct Payments
DDA	Doha Development Agenda
EFR247	EU-Wide Flat Rate (€247 per hectare)
EFR100	EU-wide Flat Rate (€100 per hectare)
EU	European Union
EU10	Member States of the EU that acceded in 2004.
EU12	Member States of the EU that acceded in 2004 and 2007
EU15	Member States of the EU that acceded prior to 2004
EU27	All 27 Member States of the EU
GAMS	General Algebraic Modelling System
GATT	General Agreement on Tariffs and Trade
GHG	Greenhouse gases
GMM	Generalized Method of Moments
GNI	Gross National Income
IPCC	Intergovernmental Panel on Climate Change
IPTS	Institute for Prospective Technological Studies
JRC	Commission's Joint Research Centre
LS	Least Squares
Mercosur	Mercado Común del Sur
ML	Maximum Likelihood
OECD	Organisation for Economic Cooperation and Development
OLS	Ordinary Least Squares
PE	Partial Equilibrium
Pillar 1	CAP policy measures directed at agricultural market and agricultural income support
Pillar 2	CAP policy measures directed at rural development
RFR	Regional Flat Rate
SADC	Southern African Development Community
SAPS	Single Area Payments Scheme

F. Chantreuil et al. (eds.), *The Future of EU Agricultural Markets by AGMEMOD*,
DOI 10.1007/978-94-007-2291-0, © Springer Science+Business Media B.V. 2012

SMP	Skim Milk Powder
SPS	Single Payment System
UAA	Utilized Agricultural Area
WTO	World Trade Organization
ZeroSPS	Zero Single Payment Support

Authors and Contributors

Frédéric Chantreuil was senior researcher at INRA for the whole duration of the fifth and sixth EU Framework Program AGMEMOD projects and moved recently to the department of economics at the Universtiy of Caen, where he graduated in Economics in 1992. He was Assistant Professor at the University of Cergy-Pontoise from 1993 to 1997. He has been the coordinator of the AGMEMOD Partnership from 2006 to 2009. His major fields of research are cooperative game theory, inequality theory and agricultural economics.

Trevor Donnellan is a Principal Research Officer at the Rural Economy Research Centre (Teagasc) in Ireland. He previously worked at the Department of Economics, University College Cork where he was part of the research team which built a survey based regional input-output model of the South West region of Ireland. He was part of the coordinating team which established the AGMEMOD commodity modelling network in 2001.

Emil Erjavec is full Professor at the Chair for agricultural economics policy and law of the University of Ljubljana and visiting Professor at the University of Natural Resources and Applied Live Sciences in Vienna. He is a graduate of these two Universities in agricultural sciences. His major fields of research are Common Agricultural Policy, European integration and agricultural sector modelling. He was the negotiator for Slovenia's accession to the EU.

Roberto Esposti is Director of Centro Interuniversitario SPERA and Senior Researcher and Assistant Professor at the Department of Economics, Università Politecnica delle Marche at Ancona. His major fields of research are environmental and agricultural economics, economics and econometrics of technical change, agricultural biotechnology and rural development analysis.

Kevin F. Hanrahan is a Principal Research Officer at the Rural Economy Research Centre (Teagasc) in Ireland. He is a graduate of Trinity College Dublin and the University of Missouri-Columbia. His research projects focus on PE modelling of agricultural commodity markets and on the quantitative analysis of the impact of agricultural policies on the agricultural sector.

F. Chantreuil et al. (eds.), *The Future of EU Agricultural Markets by AGMEMOD*,
DOI 10.1007/978-94-007-2291-0, © Springer Science+Business Media B.V. 2012

Petra Salamon is senior researcher at the Johann Heinrich von Thünen-Institute (vTI), the Federal Research Institute for Rural Areas, Forestry and Fisheries in Germany. She graduated in agricultural science and has a PhD from the Justus Liebig University of Giessen. Originally a dairy market specialist, she has been involved in market modelling using both general and partial equilibrium models. Her major fields of research include quantitative analysis and model based impact analysis. In this area she contributes to the vTI model based advisory group of the Federal Ministry of Food, Agriculture and Consumer Protection in Germany.

Guna Salputra is Head of the Unit of Sector Economics and Policy at the Latvian State Institute of Agricultural Economics. Her responsibilities within LSIAE are the assessment of policy impacts on sector development, quantitative analysis of economic development of Latvian agricultural sector and mathematical modelling.

Andrzej Tabeau is a researcher at LEI Wageningen UR in the Netherlands. He graduated and obtained his Ph.D. and Habilitation in the area of Economics from the Warsaw School of Economics in Poland. His major fields of research include quantitative analysis and future scenario development based on CGE and partial equilibrium models.

Myrna van Leeuwen is a senior researcher at LEI Wageningen UR in the Netherlands. She graduated at the Erasmus University in Rotterdam. Her research projects at LEI focus on agricultural (regional) input-output and SAM modelling, and CGE and PE agricultural modelling. Research activities are in the field of quantitative analyses of agricultural policy. She undertook the coordination of all midterm analysis of the AGMEMOD Partnership as part of collaborations with DG JRC.

The results presented in the book have been achieved through the combined efforts of women and men belonging to 24 teams all over Europe, who developed their respective country models and participated in the building of the AGMEMOD model. Therefore, the entire second part of this book, which emphasizes the demonstration of potential applications of the AGMEMOD model, is based on the knowledge and contribution of all AGMEMOD partners.

Index

F. Chantreuil et al. (eds.), *The Future of EU Agricultural Markets by AGMEMOD*,
DOI 10.1007/978-94-007-2291-0, © Springer Science+Business Media B.V. 2012

The manufacturer's authorised representative in the EU is Springer Nature Customer Service Centre GmbH, Europaplatz 3, 69115 Heidelberg, Germany. If you have any concerns regarding our products, please contact ProductSafety@springernature.com

Printed and bound by CPI Group (UK) Ltd, Croydon, CR0 4YY
30/09/2026
02225471-0009